Moving Boundary Problems

ACADEMIC PRESS RAPID MANUSCRIPT REPRODUCTION

Proceedings of the Symposium and
Workshop on Moving Boundary Problems held
at Gatlinburg, Tennessee on September 26–28, 1977.

Moving Boundary Problems

Edited by

D. G. Wilson

Alan D. Solomon

Mathematics and Statistics
Research Department
Computer Sciences Division
Union Carbide Corporation
Nuclear Division
Oak Ridge, Tennessee

Paul T. Boggs

U.S. Army Research Office
Research Triangle Park, North Carolina

ACADEMIC PRESS NEW YORK SAN FRANCISCO LONDON 1978

A Subsidiary of Harcourt Brace Jovanovich, Publishers

ACADEMIC PRESS, INC.
111 Fifth Avenue, New York, New York 10003

United Kingdom Edition published by
ACADEMIC PRESS, INC. (LONDON) LTD.
24/28 Oval Road, London NW1 7DX

Library of Congress Cataloging in Publication Data

Main entry under title:

Moving boundary problems.

Proceedings of a symposium sponsored by the Army Research Office, and the Mathematics and Statistics Research Dept. of the Computer Sciences Division of Union Carbide Corporation, Nuclear Division, held Sept. 26–28, 1977, in Gatlinburg, Tenn.

Includes indexes.

1. Boundary value problems—Congresses. I. Wilson, David George. II. Solomon, Alan D. III. Boggs, Paul T. IV. United States. Army Research Office. V. Union Carbide Corporation. Mathematics and Statistics Research Dept.

QA379.M68 515'.35 78-5125
ISBN 0-12-757350-X

PRINTED IN THE UNITED STATES OF AMERICA

Contents

List of Contributors

Numbers in parentheses indicate the pages on which the authors' contributions begin.

D. A. Antoniadis (233), Integrated Circuits Laboratory, Stanford Electronics Laboratories, Stanford University, Stanford, California

Bruno A. Boley (205), Technological Institute, Northwestern University, Evanston, Illinois

J. R. Cannon (3), Department of Mathematics, The University of Texas, Austin, Texas

Colin W. Cryer (91), Mathematics Research Center, University of Wisconsin, Madison, Wisconsin

R. W. Dutton (233), Integrated Circuits Laboratory, Stanford Electronics Laboratories, Stanford University, Stanford, California

George J. Fix (109), Department of Mathematics, Carnegie-Mellon University, Pittsburgh, Pennsylvania

Bernard A. Fleishman (41), Department of Mathematical Sciences, Rensselaer Polytechnic Institute, Troy, New York

Avner Friedman (25), Department of Mathematics, Northwestern University, Evanston, Illinois

Ross Gingrich (41), Department of Mathematical Sciences, Rensselaer Polytechnic Institute, Troy, New York

Gabriel Horvay (249), Department of Civil Engineering, University of Massachusetts, Amherst, Massachusetts

N. Kikuchi (147), Texas Institute for Computational Mechanics, The University of Texas, Austin, Texas

David Kinderlehrer (57), School of Mathematics, University of Minnesota, Minneapolis, Minnesota

Thomas J. Mahar (41), Department of Mathematics, Utah State University, Logan, Utah

Michael F. Malone (249), Department of Chemical Engineering, University of Massachusetts, Amherst, Massachusetts

Gunter H. Meyer (73), School of Mathematics, Georgia Institute of Technology, Atlanta, Georgia

Louis Nirenberg (57), Courant Institute, New York University, New York, New York

J. R. Ockendon (129), Oxford University Computing Laboratory, Oxford, England

J. T. Oden (147), Texas Institute for Computational Mechanics, The University of Texas, Austin, Texas

N. Shamsundar (165), Department of Mechanical Engineering, University of Houston, Houston, Texas

A. D. Solomon (187), Union Carbide Corporation, Nuclear Division, Oak Ridge, Tennessee

J. A. Wheeler (267), Exxon Production Research Company, Houston, Texas

Yin-Chao Yen (285), Physical Sciences Branch, Research Division, U. S. Army Cold Regions Research and Engineering Laboratory, Hanover, New Hampshire

Preface

Several years ago, as a result of a suggestion that our department sponsor occasional symposia, I proposed holding a meeting on moving boundary problems. Due to difficulties in obtaining sufficient funding, we were unable to proceed until a commonality of interest with the Army Research Office was identified. As it happened they were supporting work on moving boundary problems and were considering sponsoring a workshop. This volume contains the proceedings of the resulting symposium/workshop which was held in Gatlinburg, Tennessee, September 26–28, 1977.

The organizers were Paul Boggs of ARO, and Alan Solomon, Bob Ward, and myself from MSRD. Alan and I recruited speakers from universities and industry and spoke ourselves. Paul recruited speakers from Army installations and universities, and organized the problem session. Bob also recruited speakers, coordinated local arrangements, and served as symposium chairman. The selection of participants for the contributed paper sessions and the poster session was a communal effort.

The program arranged by the four of us, with helpful suggestions from John Cannon, Gunter Meyer, E. M. Sparrow, and others, consisted of invited talks on theory, methods, and applications; a problem session; contributed papers; a poster display maintained throughout the meeting; and a panel discussion to sum up. Of the invited talks it was intended that one in each area would be a survey or overview followed by reports of individuals' work. These proceedings are an approximate record of these invited presentations plus a summary of the panel discussion. The survey speakers' contributions begin each section of the volume. Avner Friedman and John Cannon agreed to divide the survey of the state of the theory between them. John was given precedence by virtue of abecedary.

Two speakers are not represented by written contributions. Regrettably, I am one of these. I described an ADI finite difference scheme using a weak solution approach for a two dimensional single phase Stefan problem with a boundary condition given by a linear relation between the temperature and the integral over the boundary of the flux. At the time I spoke the computer programs that implemented this scheme were not completely debugged. Although it is probably true that large programs are never completely debugged, I am not yet satisfied with these. Unfortunately, some of the impetus for solving this problem has disappeared since the engineering application from which it came has changed.

However, I have made a contribution to the proceedings which I believe will significantly increase its usefulness. I have constructed an author index and a subject index. Because these indexes are handmade and not computer cross indexed, there are doubtless omissions and other shortcomings. On the other hand, I have referenced ideas instead of just words, which computers cannot yet do. It was an educational, though somewhat tiresome, experience. I hope that it will be appreciated.

The other speaker not represented is Carl Nelson from the U.S. Army Ballistic Research Laboratory. He spoke about solid propellent combustion during the pressure transient of a gun chamber. I am reluctant to attempt a more comprehensive description of his presentation. At the meeting he indicated to me that his results would be published elsewhere.

Aside from the indexes which I prepared and the summary of the panel discussion which Alan contributed, this volume is a photoreproduction of the manuscripts the authors submitted. We took the liberty of correcting a few obvious typographical errors and filling in a few omitted references, but no extensive editing has been done.

To my personal knowledge this is only the third volume to be published on these problems. No major breakthroughs are presented. Hopefully, this is a summary of the state of the art as of September 1977.

I would like to acknowledge support from the following agencies which made the conference possible: The Army Research Office and the Mathematics and Statistics Research Department of the Computer Sciences Division of Union Carbide Corporation, Nuclear Division under contract with the United States Department of Energy.

D. G. Wilson

Moving Boundary Problems

Theory Papers

MOVING BOUNDARY PROBLEMS

MULTIPHASE PARABOLIC FREE BOUNDARY VALUE PROBLEMS

J. R. Cannon

A survey of some of the recent results for multiphase Stefan problems, chemical reaction problems, and Muskat problems is presented.

1. INTRODUCTION

In regular parabolic boundary-initial value problems, the domain of the solution of the parabolic differential equation is known along with the initial and boundary data. There is a vast literature treating such problems. For example the reader is referred to Friedman [26], Ladyzenskaja, Solonnikov, and Ural'ceva [39], and Lions [40].

The fundamental difference between regular parabolic boundary-initial value problems and parabolic free boundary value problems is that the domain of the solution of the parabolic equation is unknown and must be determined along with the solution of the parabolic equation. This requires additional information relating the solution of the parabolic equation to its domain of definition. Usually such a relationship arises naturally from the underlying physics of the particular problem under consideration.

Parabolic free boundary problems involving only one parabolic differential equation and the determination of the domain of its solution are called one phase free boundary value problems. Multiphase parabolic free boundary value problems involve several different parabolic equations defined in several different unknown domains and the determination of each domain and the solution of the corresponding parabolic equation in that domain. Here, we do not regard a collection of independent parabolic free

ISBN 0-12-757350-X

boundary value problems as a multiphase problem. The important ingredient of a multiphase problem is that each domain and the solution of the corresponding parabolic equation is linked to every other domain and corresponding solution through a set of relations that arise naturally from the physics of the problem being considered. Probably the simplest example of a multiphase parabolic free boundary value problem is the Stefan problem which attempts to model problems like a collection of ice cubes in a glass of water. The water constitutes one domain with several boundary components which bound domains representing the ice cubes. A standard way of modeling the heat flow involves one parabolic equation in the water and another in each ice cube. The solutions and the various domains are linked together through energy flow rate balances at the ice-water interfaces which involve the latent heat of fusion and the rate at which ice is converted into water (i.e. rate of change of the domains) and Fourier's law on each side of the ice-water interface.

In this article the Stefan problem will be considered first. Several recent results will be discussed. Then, a chemical reaction problem will be described and various results for it will be discussed. Finally, variations of the Muskat problem which attempts to describe the movement of immiscible incompressible fluids in porous media will be presented. The article is concluded with comments on some recent results on numerical methods.

2. STEFAN PROBLEM

A. ONE SPACE VARIABLE

Given $T > 0$, $b \in (0,1)$, $f_1 = f_1(t) > 0$ $(0 \le t \le T)$, $f_2 = f_2(t) < 0$ $(0 \le t \le T)$, and $\varphi = \varphi(x)$ $(0 \le x \le 1)$ such that $(b-x)\varphi(x) \ge 0$ $(0 \le x \le 1)$, determine a curve $x = s(t)$ $(0 \le t \le T)$, which divides the rectangular domain

$$Q_T = \{(x,t);\ 0 < x < 1,\ 0 < t \le T\}$$

into two subdomains

$$D_T^{\pm} = \{(x,t) \in Q_T;\ \pm (x-s(t)) > 0\},$$

and a function $u = u(x,t)$, defined and continuous on $\bar{Q}_T$, such that

$$0 < s(t) < 1, \quad s(0) = b, \quad s \in C([0,T]) \cap C^1((0,T]); \tag{2.1}$$

$$u \in C^{2,1}(D_T^-) \quad \text{and} \quad \frac{\partial u}{\partial t} = \beta_1 \frac{\partial^2 u}{\partial x^2} \quad \text{in } D_T^-, \tag{2.2}$$

$$u \in C^{2,1}(D_T^+) \quad \text{and} \quad \frac{\partial u}{\partial t} = \beta_2 \frac{\partial^2 u}{\partial x^2} \quad \text{in } D_T^+, \tag{2.3}$$

$$\begin{aligned} u(0,t) &= f_1(t), \quad 0 \le t \le T, \\ u(1,t) &= f_2(t), \quad 0 \le t \le T, \end{aligned} \tag{2.4}$$

$$\left.\begin{aligned} u(x,0) &= \varphi(x), \quad 0 \le x \le b, \\ u(x,0) &= \varphi(x), \quad b \le x \le 1, \end{aligned}\right\} \tag{2.5}$$

$$u(s(t),t) = 0, \quad 0 \le t \le T, \tag{2.6}$$

$$\alpha \dot{s}(t) = -\gamma_1 \frac{\partial u}{\partial x}(s(t)-,t) + \gamma_2 \frac{\partial u}{\partial x}(s(t)+,t), \quad 0 < t \le T. \tag{2.7}$$

The quantities $\beta_1, \beta_2, \gamma_1, \gamma_2$ and α involve the positive physical parameters of the problem.

The system (2.1)-(2.7) is a simple model of a heat-conducting substance consisting of two phases (e.g. liquid and solid) of the same substance which are in perfect thermal contact at an interface. The function $u(x,t)$ represents the temperature distribution in the system, and the curve $x = s(t)$ represents the position of the interface, which varies with time t as solid melts or liquid freezes. The boundary conditions at the interface ((2.6) and (2.7)) reflect respectively the facts that the temperature at the interface must be equal to the melting temperature (taken to be zero) and that the rate of melting is proportional to the rate of absorption of heat energy at the interface.

An excellent historical survey of problems of this type is provided by the monograph of Rubinstein [43]. Since 1967 (the date of publication of [43]), numerous advances in this area have taken place. Cannon and Primicerio [8,9,10,11], following the work of Cannon, Douglas and Hill [6], showed that this problem has a unique classical solution provided that the f_i and φ are bounded by certain constants which depend on the parameters α, β_1, β_2, γ_1 and γ_2. Friedman [27,28] developed a theory of weak solutions of Stefan problems which in the case of (2.1)-(2.7) [28] yields an $s(t)$ which is continuous and satisfies (2.7) in a generalized sense.

For (2.1)-(2.7) with $\gamma_1 = \gamma_2 = 1$, Friedman's definition of weak solution is as follows:

Definition. A bounded measurable function $u(x,t)$ in $\Omega_T = \{(x,t);\ 0 < t \leq T,\ 0 < x < 1\}$ is called a weak solution if, for all $\psi \in C^{2,1}(\bar{\Omega}_T)$ such that $\psi = 0$ for $x = 0,1$ and $t = T$, we have

$$\int_0^T \int_0^1 [u\psi_{xx} + a(u(x,t))\psi_t]\,dx\,dt = -\int_0^1 a(\varphi(x))\psi(x,0)\,dx$$
$$+ \int_0^T [f_2(t)\psi_x(1,t) - f_1(t)\psi_x(0,t)]\,dt$$

where

$$a(u(x,t)) = \begin{cases} \beta_1 u(x,t) & \text{where } u(x,t) > 0 \\ -\alpha + \beta_2 u(x,t) & \text{where } u(x,t) < 0 \\ \text{some measurable function of } (x,t) \text{ with} \\ \text{range} \subseteq [-\alpha, 0] & \text{where } u(x,t) = 0. \end{cases}$$

The results of Friedman are summarized in the following

Theorem 1. (A. Friedman [28]). Assume that $\varphi' \in L_2(0,1)$ and that there exists $\Psi(x,t)$ in $\bar{\Omega}_T$ with Ψ_x, Ψ_{xx}, Ψ_t continuous in $\bar{\Omega}_T$ and $\Psi(0,t) = f_1(t)$, $\Psi(1,t) = f_2(t)$, $\Psi(x,0) = \varphi(x)$ for $0 \le t \le T$ and x close to 0 or 1. Then

(a) There exists a unique weak solution $u(x,t)$.

(b) $u(x,t)$ is continuous on $\bar{\Omega}_T$, smooth in $\Omega_T = \{u = 0\}$, and satisfies $u_t = \beta_1 u_{xx} [u_t = \beta_2 u_{xx}]$ in $\Omega_T \cap \{u > 0\}$ [resp. $\Omega_T \cap \{u < 0\}$].

(c) There exists $d > 0$, which depends only on the data, such that $u > 0$ $[u < 0]$ for $|x| \le d$ [resp. $|1-x| \le d$].

(d) There exists a constant C such that

$$\left[\int_0^1 |u_x(x,t)|^2 dx\right]^{1/2} \le C \quad (0 \le t \le T).$$

(e) For each $t \in [0,T]$ and x_1, $x_2 \in [0,1]$,

$$|u(x_1,t) - u(x_2,t)| \le C|x_1 - x_2|^{1/2}.$$

(f) For each $t \in [0,T]$, there exists a unique $s(t)$ such that $u(s(t),t) = 0$.

(g) $s(t)$ is a continuous function of t for $0 \le t \le T$.

(h) The energy relation

$$\alpha[s(t)-b] = \beta_1^{-1} \int_0^b \varphi(x)dx + \beta_2^{-1} \int_b^1 \varphi(x)dx - \beta_1^{-1} \int_0^{s(t)} u(x,t)dx - \beta_2^{-1} \int_{s(t)}^1 u(x,t)dx + \int_0^t [u_x(1,\tau) - u_x(0,\tau)]d\tau \tag{2.8}$$

is satisfied for each $t \in [0,T]$.

Cannon, Henry and Kotlow [13] reported the following result. Some of the proofs which were omitted in [13] are supplied here.

Theorem 2. (Cannon, Henry, and Kotlow [13]). Let $u(x,t)$ and $s(t)$ satisfy (b)-(h) of Theorem 1, and assume (without loss of generality) that $s(t)$ is continuous for $0 \leq t \leq T$. Then $s(t)$ is continuously differentiable for $0 < t \leq T$, $u_x(s(t) \pm 0, t)$ are well-defined, bounded and continuous for $0 < t \leq T$ and

$$\alpha \dot{s}(t) = -u_x(s(t) - 0, t) + u_x(s(t) + 0, t) \quad (0 < t \leq T). \tag{2.9}$$

In particular, Theorems 1 and 2 guarantee the existence of a unique classical solution without the size restrictions on the data which were imposed in [6,8,9,10,11].

The first step in the proof of Theorem 2 is to show that $s(t)$ satisfies a Holder condition with exponent 3/4.

Lemma. (Cannon, Henry, and Kotlow [13]). Let $u(x,t)$ and $s(t)$ be as in Theorem 2. Then there exists a constant M such that

$$|s(t_1) - s(t_2)| \leq M|t_1 - t_2|^{3/4} \quad (0 \leq t_1, t_2 \leq T). \tag{2.10}$$

Proof. For any $\eta \in (0,T]$, let $m(\eta)$ denote the maximum oscillation of $s(t)$ over intervals of length η. It is sufficient to show that $m(\eta) \leq M\eta^{3/4}$ for all $\eta \in (0,\eta_0]$, where $\eta_0 > 0$. Since $s(t)$ is uniformly continuous on $[0,T]$, it is possible to choose η_0 such that

$$\eta_0 < d^2/16, \; m(\eta) < \min[d/4, 1/4C_0^2] \; (0 < \eta < \eta_0), \tag{2.11}$$

where C_0 is a constant to be specified later and d is as in Theorem 1(c). Consider below $\eta < \eta_0$ as fixed.

Choose $t_1 \in [0,T]$ such that $|s(t_1 + \theta\eta) - s(t_1)| = m(\eta)$ for some $\theta \in [0,1]$; it is no loss of generality to suppose that $\theta = 1$ and $s(t_1 + \eta) = s(t_1) - m(\eta)$. For any $\delta \in (0, d/4)$, the rectangle $R(\eta,\delta) = \{(x,t);\ s(t_1) - m(\eta) - \delta \le x \le s(t_1) + m(\eta) + \delta,\ t_1 \le t \le t_1 + \eta\}$ lies in $\bar{\Omega}_T$, and the curve $x = s(t)$ does not intersect the edges which are parallel to the t-axis. It is easy to localize the energy relation (2.8) to $R(\eta,\delta)$; setting $\Delta = m(\eta) + \delta$, we have

$$
\begin{aligned}
\alpha[s(t_1+\eta) - s(t_1)] = \beta_1^{-1} \int_{s(t_1)-\Delta}^{s(t_1)} u(x,t_1)dx \\
+ \beta_2^{-1} \int_{s(t_1)}^{s(t_1)+\Delta} u(x,t_1)dx \\
- \beta_1^{-1} \int_{s(t_1)-\Delta}^{s(t_1+\eta)} u(x,t_1+\eta)dx \\
- \beta_2^{-1} \int_{s(t_1+\eta)}^{s(t_1)+\Delta} u(x,t_1+\eta)dx \\
+ \int_{t_1}^{t_1+\eta} [u_x(s(t_1)+\Delta,t) - u_x(s(t_1)-\Delta,t)]dt.
\end{aligned}
\tag{2.12}
$$

By virtue of Theorem 1(e) and (f), the first four terms on the right side of (2.12) may be estimated by const. $\Delta^{3/2}$. Since the left side is equal to $-\alpha m(\eta)$, it follows that

$$
\begin{aligned}
m(\eta) \le C\{\Delta^{3/2} + \int_{t_1}^{t_1+\eta} [\,|u_x(s(t_1)+\Delta,t)| \\
+ |u_x(s(t_1)-\Delta,t)|\,]dt\}.
\end{aligned}
\tag{2.13}
$$

Both sides of (2.13) can be averaged with respect to δ from $\delta = 0$ to $\delta = \delta_0 < d/4$. Since

$$\delta_0^{-1} \int_{t_1}^{t_1+\eta} \int_0^{\delta_0} [|u_x(s(t_1)+\Delta,t)| + |u_x(s(t_1)-\Delta,t)|] d\delta\, dt$$

$$\leq 2\delta_0^{-1}\eta\left(\int_0^{\delta_0} 1\, d\delta\right)^{1/2} \cdot \sup_t \left(\int_0^1 |u_x(x,t)|^2 dx\right)^{1/2}$$

$$\leq C\delta_0^{-1/2}\eta$$

by Theorem 1(d) and Schwartz' inequality, one obtains the result

$$m(\eta) \leq C[\Delta_0^{3/2} + \delta_0^{-1/2}\eta] \leq C_0[m(\eta)^{3/2} + \delta_0^{3/2} + \delta_0^{-1/2}\eta], \qquad (2.14)$$

where $\Delta_0 = m(\eta) + \delta_0$. Now, set $\delta_0 = \eta^{1/2}$ in (2.14). Taking (2.11) into account, $\delta_0 < d/4$ as required, and also $C_0 m(\eta)^{3/2} \leq m(\eta)/2$, so that (2.14) yields $m(\eta) \leq 4C_0\eta^{3/4}$. ●

The Lemma enables the utilization of the following result concerning the boundary behavior of the derivative of a solution of the heat equation which vanishes on a non-smooth boundary curve.

Theorem 3. (Cannon, Henry and Kotlow [14]). Let $s(t)$ be such that $s(t) \geq d > 0$ $(0 \leq t \leq T)$, $s(0) = b$ and

$$|s(t_1) - s(t_2)| \leq M|t_1 - t_2|^{\lambda} \quad (0 \leq t_1, t_2 \leq T)$$

where $1/2 < \lambda \leq 1$. Let $v(x,t)$ be the solution of the problem (i) $v_t = v_{xx}$ $(0 < x < s(t),\ 0 < t \leq T)$; (ii) $v(x,0) = \phi(x)$ $(0 \leq x \leq b)$; (iii) $v(0,t) = f(t)$ $(0 \leq t \leq T)$; (iv) $v(s(t),t) = 0$ $(0 \leq t \leq T)$; where $f(t)$ and $\phi(x)$ are continuous with

$f(0) = \phi(0)$ and $|\phi(x)| \leq K(b-x)$ $(0 \leq x \leq b)$. Then $v_x(x,t)$ converges as $x \to (s(t) - 0, t)$ to a limit $v_x(s(t) - 0, t)$ which is a bounded continuous function of t for $0 < t \leq T$. Moreover, the convergence is uniform on $[\delta, T]$ for any $\delta > 0$.

Since the proof is quite long and technical, the reader is referred to [14].

To complete the proof of Theorem 2, it is sufficient to show that

$$\alpha[s(t_2) - s(t_1)] = \int_{t_1}^{t_2} [-u_x(s(t) - 0, t) + u_x(s(t) + 0, t)]dt \tag{2.15}$$

for any t_1, $t_2 \in (0,T]$. To this end, let $t_1 < t_2$ be fixed, let $\eta > 0$ be an arbitrary positive number, let $\varepsilon = M[(t_2 - t_1)/N]^{3/4}$, and choose an integer $N > (2C_1M/\eta)^8 (t_2 - t_1)^9$, where C_1 is a constant to be specified later. Also, take N so large that

$$|u_x(x,t) - u_x(s(t) \pm 0, t)| < [4(t_2 - t_1)]^{-1}\eta$$

whenever $t_1 \leq t \leq t_2$ and $0 < \pm(x - s(t)) < 2\varepsilon$ (this is possible by virtue of Theorem 3 and the Lemma). Now choose a partition $t_1 = \tau_0 < \tau_1 \cdots < \tau_N = t_2$ with $\tau_{i+1} - \tau_i = (t_2 - t_1)/N$ $(i = 0, \ldots, N-1)$. By (3) we have $s(\tau_i) - \varepsilon \leq s(t) \leq s(\tau_i) + \varepsilon$ for $t \in [\tau_i, \tau_{i+1}]$ $(i = 0, \ldots, N-1)$. Write down the energy relations in the rectangles $R_i = \{(x,t);\ s(\tau_i) - \varepsilon \leq x \leq s(\tau_i) + \varepsilon, \tau_i \leq t \leq \tau_{i+1}\}$ and sum from $i = 0$ to $i = N - 1$; the result is

$$\alpha[s(t_2) - s(t_1)] = \sum_{i=0}^{N-1} [\beta_1^{-1} \int_{s(\tau_i)-\varepsilon}^{s(\tau_i)} u(x,\tau_i)dx$$

$$+ \beta_2^{-1} \int_{s(\tau_i)}^{s(\tau_i)+\varepsilon} u(x,\tau_i)dx$$

$$- \beta_1^{-1} \int_{s(\tau_i)-\varepsilon}^{s(\tau_{i+1})} u(x,\tau_{i+1})dx - \beta_2^{-1} \int_{s(\tau_{i+1})}^{s(\tau_i)+\varepsilon} u(x,\tau_{i+1})dx]$$

$$+ \sum_{i=0}^{N-1} \int_{\tau_i}^{\tau_{i+1}} [u_x(s(\tau_i)+\varepsilon,t) - u_x(s(\tau_i)-\varepsilon,t)]dt.$$

By Theorem 1(e), each of the integrals under the first summation sign may be estimated by const. $\varepsilon^{3/2}$. Hence,

$$|\alpha[s(t_2) - s(t_1)] - \int_{t_1}^{t_2} [-u_x(s(t)-0,t) + u_x(s(t)+0,t)]dt|$$

$$\leq C_1 N\varepsilon^{3/2} + \sum_{i=0}^{N-1} \sum_{j=1}^{2} \int_{\tau_i}^{\tau_{i+1}} |u_x(s(\tau_i) + (-1)^j\varepsilon, t) - u_x(s(t) + (-1)^j 0, t)|dt$$

$$\leq C_1 M(t_2 - t_1)^{9/8} N^{-1/8} + 2(t_2 - t_1)\cdot[4(t_2 - t_1)]^{-1}\eta < \eta,$$

which proves (2.15). ●

In [14], Cannon, Henry and Kotlow demonstrated the following result for (2.1)-(2.7) via potential theoretic techniques.

Theorem 4. (Cannon, Henry and Kotlow [14]). If $f_i \in C^1([0,T])$, $i = 1,2$, $(-1)^{i-1} f_i(t) \geq m > 0$, $0 \leq t \leq T$, $\varphi \in H^1([0,1])$, then there exist $s \in C^{3/4}([0,T]) \cap C^1((0,T])$ and u, continuous on $\overline{Q}_T$, satisfying the conditions (2.1)-(2.7) of the problem. Also if $\varphi \in C^\beta([0,1])$ for $\beta \in [1/2,1]$ ($\beta = 1/2$ already satisfied above), then s and s are Holder continuous with exponent depending upon β.

Fasano and Primicerio [24] have recently given another demonstration of the regularity of weak solutions of (2.1)-(2.7). Their proof utilizes the results of Schaefer [44].

Theorem 5. (Schaefer [44]). For (2.1)-(2.7) if $\dot{s}$ is Holder continuous for $t > 0$, then $s \in C_\infty((0,T])$ and all derivatives of u are continuous up to the free boundary curve $x = s(t)$.

Recently, Friedman has demonstrated the following result

Theorem 6. (Friedman [30]). For (2.1)-(2.7) if f_i, $i = 1,2$ are analytic, then s is analytic.

Many modifications of (2.1)-(2.2) are possible. In [9,10,11] Cannon and Primicerio considered $b = 0$ and flux boundary conditions. Hill and Kotlow [34,35] replaced (2.6) and (2.7) with the conditions

$$\begin{aligned} & u(s(t)+,t) = 0, \qquad 0 \le t \le T \\ & -\gamma_1 \frac{\partial u}{\partial x}(s(t)-,t) = hu(s(t)-,t), \qquad 0 < t \le T, \\ & [\alpha + \sigma u(s(t)-,t)]\dot{s}(t) = -\gamma_1 \frac{\partial u}{\partial x}(s(t)-,t) + \\ & \qquad + \gamma_2 \frac{\partial u}{\partial x}(s(t)+,t), \qquad 0 < t \le T, \end{aligned} \tag{2.16}$$

where σ and h are positive constants. Boundary conditions including the flux depending non-linearly upon the boundary temperature were considered. Fasano and Primicerio [21,22,23,25] have recently concluded a study in which (2.6) and (2.7) are replaced with

$$u(s(t),t) = f(s(t),t), \qquad 0 < t \leq T,$$
$$\dot{s}(t) = F\left(t, s(t), \frac{\partial u}{\partial x}(s(t)-,t), \frac{\partial u}{\partial x}(s(t)+,t)\right), \quad 0 < t \leq T, \tag{2.17}$$

where F depends linearly upon $\frac{\partial u}{\partial x}(s(t)-,t)$ and $\frac{\partial u}{\partial x}(s(t)+,t)$. Various types of boundary conditions are considered. Very recently, Evans and Kotlow [18] have demonstrated global existence uniqueness, and regularity theorems for (2.1)-(2.7) in which the parabolic differential equations are quasi-linear: $\frac{\partial u}{\partial t} = \frac{\partial}{\partial x}(\beta_i(x,u)\frac{\partial u}{\partial x})$ and the $\gamma_i = \beta_i(s(t),0)$.

B. SEVERAL SPACE VARIABLES

The description of this problem is similar to that for the chemical reaction problem and is deferred to section 3. The results for the Stefan problem will be included there as a subset of that discussion.

3. A CHEMICAL PROBLEM

Suppose that two substances in solution diffuse and when they come together at Γ (a set implicitly defined here) they chemically react quickly and completely. Hence at Γ the concentrations are zero. Chemically, a certain amount of substance 1 must flow in from one side of Γ in order to react with a proportional amount of substance 2 flowing in from the opposite side of Γ. Similar to the Stefan problem (see Friedman [27]) in several variables, one cannot expect Γ to have a simple representation. Hence, a weak formulation is natural In order to motivate it, one has to work through a classical setting which assumes a lot of regularity on Γ.

In [7], Cannon and Hill, utilized Friedman's [27] formulation of the Stefan problem in several variables to derive the following description of the chemical problem. Let G be a bounded domain in R^n with a smooth boundary ∂G. Set $G(t) = G \times \{t\}$ and $\partial G(t) = \Omega_T = \cup_{0<t<T} G(t)$ and $S_T = \cup_{0<t<T} \partial G(t)$. When there is

no danger of confusion T can be dropped and Ω and S can be employed for Ω_T and S_T. Ω is divided into two parts, Ω_1 and Ω_2, by the free boundary $\Gamma = \Gamma_T = \bigcup_{0\leq t\leq T}\Gamma(t)$, where $\Gamma(t)$ is a hypersurface in $\overline{G(t)}$ given by $\Gamma(t) = \{(x,t) \in \overline{G(t)} \mid \Phi(x,t) = 0\}$ and where $\Phi \in C^1(\overline{\Omega})$ and $\nabla_x\Phi(x,t) \neq 0$ on Γ. It can be assumed that $\Phi < 0$ in Ω_1 and $\Phi > 0$ in Ω_2. In fact $\Gamma(0)$ divides the initial region $G(0)$ into two regions $G_1(0)$ and $G_2(0)$. Let $S_i = \overline{\Omega}_i \cap S$, $i = 1,2$, and note that some S_i might be void

Given the data $\{\tilde{h}_1, \tilde{h}_2, \Gamma(0)\}$ the problem is to find a solution $\{\tilde{u}_1, \tilde{u}_2, \Phi\}$ which satisfies

$$\alpha_i(\tilde{u}_i)\frac{\partial \tilde{u}_i}{\partial t} = \sum_{j=1}^{n} \frac{\partial}{\partial x_j}[k_i(\tilde{u}_i)\frac{\partial \tilde{u}_i}{\partial x_j}] \quad \text{in } \Omega_i, \quad i = 1,2 \tag{3.1}$$

$$\tilde{u}_i = \tilde{h}_i, \quad \tilde{h}_i > 0 \quad \text{in } G_i(0), \quad i = 1,2, \tag{3.2}$$

$$k_i(\tilde{u}_i)\frac{\partial \tilde{u}_i}{\partial n} = 0 \quad \text{on } S_i, \quad i = 1,2, \tag{3.3}$$

$$\tilde{u}_i = 0 \quad \text{on } \Gamma, \quad i = 1,2, \tag{3.4}$$

$$\nu k_1(\tilde{u}_1)\sum_{j=1}^{n} \frac{\partial \Phi}{\partial x_j}\frac{\partial \tilde{u}_1}{\partial x_j} = -k_2(\tilde{u}_2)\sum_{j=1}^{n} \frac{\partial \Phi}{\partial x_j}\frac{\partial \tilde{u}_2}{\partial x_j} \quad \text{on } \Gamma, \tag{3.5}$$

where ν is a positive constant, $\frac{\partial}{\partial n}$ denotes differentiation in the direction of the outer normal to S and the $\alpha_i(u)$, $k_i(u)$ are sufficiently smooth functions defined for $u \geq 0$ which satisfy

$$0 < \gamma_0 < \alpha_i(u), \quad k_i(u) < \gamma_1, \quad i = 1,2, \tag{3.6}$$

for some constants γ_0 and γ_1.

By making even extensions of $\alpha_2(u)$ and $k_2(u)$ to negative values of u,

$$A_i(u) = \int_0^u \alpha_i(\zeta)d\zeta, \qquad i = 1,2,$$

and

$$K_i(u) = \int_0^u k_i(\zeta)d\zeta, \qquad i = 1,2,$$

can be defined so that $A_2(u)$ and $K_2(u)$ are odd functions. Setting $u_1 = \tilde{u}_1$, $h_1 = \tilde{h}_1$, $u_2 = -\tilde{u}_2$, and $h_2 = -\tilde{h}_2$, one obtains

$$\frac{\partial}{\partial t} A_i(u_i) = \text{div grad } K_i(u_i) \quad \text{in } \Omega_i, \quad i = 1, \tag{3.1'}$$

$$\begin{aligned} u_1 &= h_1 > 0 \quad \text{in } G_1(0) \\ u_2 &= h_2 < 0 \quad \text{in } G_2(0) \end{aligned} \tag{3.2'}$$

$$\frac{\partial}{\partial n} K_i(u_i) = 0 \quad \text{on } S_i, \quad i = 1,2, \tag{3.3'}$$

$$u_i = 0 \quad \text{on } \Gamma, \quad i = 1,2, \tag{3.4'}$$

$$\nu \text{ grad } \Phi \cdot \text{grad } K_1(u_1) = \text{grad } \Phi \cdot \text{grad } K_2(u_2) \quad \text{on } \Gamma \tag{3.5'}$$

where div and grad refer only to the x variables. Multiplying (3.1') by smooth test functions ϕ in R^{n+1} such that $\phi \equiv 0$ on $G(T)$ and $\frac{\partial \phi}{\partial n} \equiv 0$ on S, integrating by parts and employing the divergence theorem, multiplying the results for $i = 1$ by ν, adding them to those for $i = 2$, and defining

$$u = \begin{cases} u_1 & \text{in } \Omega_1, \\ 0 & \text{on } \Gamma, \\ u_2 & \text{in } \Omega_2, \end{cases} \qquad h = \begin{cases} h_1 > 0 & \text{in } G_1(0), \\ 0 & \text{on } \Gamma(0), \\ h_2 < 0 & \text{in } G_1(0), \end{cases}$$

$$a(u) = \begin{cases} \nu A_1(u) , & u > 0 , \\ 0 , & u = 0 , \\ A_2(u) , & u < 0 , \end{cases}$$

and

$$b(u) = \begin{cases} \nu K_1(u) , & u > 0 , \\ 0 , & u = 0 , \\ K_2(u) , & u < 0 , \end{cases}$$

one obtains

$$\int\int_\Omega [a(u)\frac{\partial\phi}{\partial t} + b(u)\Delta\phi] dx\, dt + \int_\Omega a(h)\phi dx = 0. \tag{3.7}$$

<u>Definition</u>. By a weak solution of (3.1)-(3.5) in Ω it is meant a bounded measurable function u in Ω such that (3.7) holds for all test functions satisfying $\phi \equiv 0$ in $G(T)$ and $\frac{\partial\phi}{\partial n} \equiv 0$ on S.

It should be remarked here that Friedman [27] obtained a similar weak formulation for the Stefan problem. However the $a(u)$ obtained is monotone increasing with a jump discontinuity at $u = 0$. Herein lies the difficulty with the theoretical and the numerical analysis of the Stefan problem in several variables.

Employing arguments similar to those of Friedman [27], Cannon and Hill [7] demonstrated the existence, uniqueness, monotone dependence, stability, and some regularity for weak solutions of (3.1)-(3.5).

Cannon and Fasano [12] considered two problems of the type (3.1)-(3.5). One was obtained by replacing (3.3') by $\frac{\partial}{\partial n} K_i(u_i) = \psi_i(x,t)$ on S_i, $i = 1,2$, while the other was obtained by replacing (3.3') by $K_i(u_i) = K_i(f_i(x,t))$ on S_i, $i = 1,2$. Generalizations of the arguments of Cannon and Hill [7] yielded existence, uniqueness, continuous dependence, monotone dependence and some regularity for the weak solutions of both problems.

Clearly, Friedman [27] obtained existence, uniqueness, continuous dependence monotone dependence, and some regularity for weak solutions of the Stefan problem. Brezis [2] gives a theory of even weaker solutions which applies under more general hypotheses than Friedman's theory. Henry [34] employed Brezis' theory to obtain existence and uniqueness of periodic solutions to the Stefan problem.

One of the outstanding problems in multi-dimensional free boundary problems is the characterization of the free boundary. Results in this direction have been obtained by Friedman and Kinderlehrer [31,32] and Caffarelli and Riviere [3,4,5].

Returning briefly to (3.1)-(3.5), Cannon and DiBenedetto [17] have recently considered a problem in which (3.3') was replaced by $\frac{\partial}{\partial n} K_i(u_i) = g_i(x,t,u_i(x,t))$ on S_i, $i = 1,2$. Formulating a stronger version of the generalized problem along the lines of Ladyzenskaja, Solonnikov and Ural'ceva [39], the arguments of Cannon and Ewing [16] are employed to study existence, uniqueness continuous dependence and regularity of the weak solution.

4. A GENERALIZED MUSKAT PROBLEM

A classical theory held forth by Muskat [42] was that when two immiscible fluids in a porous medium are in contact with one another an interface is formed and the movement of the fluids results in a free boundary problem. An application of the law of conservation of mass coupled with Darcy's law yields the problem of determining the pressures p,q and the interface s which as a triple (p,q,s) satisfy

$$\left.\begin{aligned}
&(1)\quad \frac{\partial\varphi(x,p)}{\partial t} = \frac{\partial}{\partial x}[a(x,p)\frac{\partial p}{\partial x}],\ 0<x<s(t),\ 0<t\leq T,\\
&(2)\quad \frac{\partial\varphi(x,q)}{\partial t} = \frac{\partial}{\partial x}[b(x,p)\frac{\partial q}{\partial x}],\ s(t)<x<1,\ 0<t\leq T,\\
&\qquad\qquad s(0)=s_0,\ 0<s_0<1,\\
&(3)\quad p(x,0) = h_1(x),\quad 0<x<s_0,\\
&(4)\quad q(x,0) = h_2(x),\quad s_0\leq x\leq 1,\\
&(5)\quad p(0,t) = f_1(t),\quad 0<t\leq T,\\
&(6)\quad q(1,t) = f_2(t),\quad 0<t\leq T,\\
&(7)\quad p(s(t),t) = q(s(t),t),\quad 0<t\leq T,\\
&(8)\quad a(s(t),p(s(t),t))\frac{\partial p}{\partial x}(s(t),t) =\\
&\qquad = b(s(t),q(s(t),t))\frac{\partial q}{\partial x}(s(t),t),\quad 0<t\leq T,
\end{aligned}\right\}\qquad (4.1)$$

and

$$\varphi(s(t),p(s(t),t))\dot{s}(t) = -a(s(t),p(s(t),t))\frac{\partial p}{\partial x}(s(t),t),\qquad 0<t\leq T,\qquad (4.2)$$

where the functions φ, a, b, h_1, h_2, f_1 and f_2 are given functions of their respective arguments and the s_0, $0 < s_0 < 1$, is a specified constant.

In [33], Fulks and Guenther considered a simpler version of (4.1)-(4.2) in which φ was a positive constant times p and a and b were positive constants. Employing potential theoretic methods, they demonstrated existence, uniqueness and continuous dependence upon the data. Cannon and Fasano [15] considered (4.1)-(4.2), derived a generalized formulation for it, and demonstrated existence of a weak solution for T sufficiently small. Recently, Evans [19,20] has considered a problem similar to the one considered by Fulks and Guenther and has demonstrated a global (large T) existence and some regularity of a weak solution of a generalized formulation of the problem.

The applicability of (4.1)-(4.2) to the movement of oil by water is limited due to the mobility ratio of oil versus that of water. Experimental evidence indicated the fingering of water into the oil causing the oil industry to shift to a mass fraction concept in the 1950's. However the possibility of other applications exists and in any event the problem (4.1)-(4.2) is mathematically interesting.

5. NUMERICAL METHODS

The reader can find many articles in this volume dealing with numerical techniques for solving various free boundary value problems. However, the recent work of some of our foreign colleagues should be reported regardless of possible duplication. Mori [41] has studied the application of the finite element technique to solving the one-phase Stefan problem. Fremond (Laboratorie Central des Ponts et Chaussees, 58 Bld Lefebvre 75732 Paris CEDEX 15) has applied the method of variational inequalities and its inherent numerical technique to the solution of "Frost propogation in wet porous media". Borgioli, DiBenedetto, and Ughi [0,1] have applied the method of Huber [37] to various Stefan problems. Hoffman (Freie Universitat Berlin, III Mathematischen Institut, Arnimallee 2-6, 1 Berlin 33, Germany) and his colleagues have just published a set of notes (preprints 22 and 28) on their various researches into the numerical solution of free boundary value problems. They have also included an extensive bibliography of the literature on free boundary value problems.

REFERENCES

0. G. Borgioli, E. DiBenedetto, and M. Ughi, Il metodo i Huber per problemi a contorno libero con dominio inizialmente degenere, preprint of Instituto di Matematico "Ulisse Dini", Viale Morgagni 67/A, 1 50134 Firenze, Italia.

1. G. Borgioli, E. DiBenedetto and M. Ughi, Stefan problems with non linear boundary conditions: The polygonal method, preprint of Istituto di Matematico "Ulisse Dini", IBIB.

2. H. Brezis, On some degenerate nonlinear parabolic equations, in Nonlinear Functional Analysis, Proc. Symp. Pure Math. vol. 18 (F. Browder, ed.) Amer. Math. Soc., Providence (1970), pp. 28-38.

3. L.A. Caffarelli, The regularity of elliptic and parabolic free boundaries, Bull. Amer. Math. Soc., vol. 82(1976), pp. 616-618.

4. L.A. Caffarelli and N.M. Riviere, On the rectifiability of domains with finite perimeter, Ann. Scuola Norm. Sup. Pisa Cl. Sci., (4) 3(1976), No. 2, pp. 177-186.

5. L.A. Caffarelli and N.M. Riviere, Smoothness and analyticity of free boundaries in variational inequalities, to appear.

6. J.R. Cannon, Jim Douglas, Jr. and C.D. Hill, A multi-boundary Stefan problem and the disappearance of phases, J. of Math. and Mech., vol. 17(1967), pp. 21-34.

7. J.R. Cannon and C.D. Hill, On the movement of a chemical reaction interface, Indiana U. Math. J., vol. 20(1970), pp. 429-454.

8. J.R. Cannon and M. Primicerio, A two phase Stefan problem with temperature boundary conditions, Ann. di Mat. pura ed appl. (IV), vol LXXXVIII (1971), pp. 177-192.

9. J.R. Cannon and M. Primicerio, A two phase Stefan problem with flux boundary conditions, Ann. di Mat. pura ed. appl (IV) vol. LXXXVIII (1971), pp. 193-206.

10. J.R. Cannon and M. Primicerio, A two phase Stefan problem: regularity of the free boundary, Ann. di Mat. pura ed appl. (IV), vol LXXXVIII (1971), pp. 217-228.

11. J.R. Cannon and M. Primicerio, A Stefan problem involving the appearance of a phase, SIAM J. Math. Anal., vol. 4 (1973), pp. 141-147.

12. J.R. Cannon and A. Fasano, Boundary value multidimensional problems in fast chemical reactions, vol. 53(1973), pp. 1-13.

13. J.R. Cannon, D.B. Henry and D.B. Kotlow, Continuous differentiability of the free boundary for weak solutions of the Stefan problem, Bull. of Amer. Math. Soc., vol. 80(1974), pp. 45-48.

14. J.R. Cannon, D.B. Henry and D.B. Kotlow, Classical solutions of the one-dimensional two-phase Stefan problem, Ann. di Mat. pura ed. appl. (IV), vol. CVII (1976), pp. 311-341.

15. J.R. Cannon and A. Fasano, A non-linear parabolic free boundary problem, Ann. di Mat. pura ed. appl. (IV), vol. CXII (1977), pp. 119-148.

16. J.R. Cannon and R.E. Ewing, Quasilinear parabolic system with non-linear boundary conditions, to appear.

17. J.R. Cannon and E. DiBenedetto, Multidimensional problems in fast chemical reactions with non-linear boundary conditions, to appear.

18. L.C. Evans and D.B. Kotlow, One-dimensional Stefan problems with quasilinear heat conduction, to appear.

19. L.C. Evans, A free boundary problem: the flow of two immiscible fluids in a one-dimensional porous medium I, to appear in the Indiana U. Math. Journal.

20. L.C. Evans, A free boundary problem: the flow of two immiscible fluids in a one-dimensional porous medium II, to appear in the Indiana U. Math. Journal.

21. A. Fasano and M. Primicerio, General free-boundary problems for the heat equation I, J. Math. Anal. and Appl., vol. 57(1977), pp. 694-723.

22. A. Fasano and M. Primicerio, General free-boundary problems for the heat equation II, J. Math. Anal. and Appl., vol. 58(1977), pp. 202-231.

23. A. Fasano and M. Primicerio, General free-boundary problems for the heat equation III, J. Math. Anal. and Appl., vol. 59(1977), pp. 1-14.

24. A. Fasano and M. Primicerio, Regularity of weak solutions of one-dimensional two-phase Stefan problems, to appear.

25. A. Fasano and M. Primicerio, Free boundary problems for nonlinear parabolic equations with nonlinear free boundary conditions, to appear.

26. A. Friedman, Partial Differential Equations of Parabolic Type, Prentice-Hall, Englewood Cliffs, N.J., 1964.

27. A. Friedman, The Stefan problem in several space variables, Trans. Amer. Math. Soc., vol. 132(1968), pp. 51-87; Correction, vol. 142(1969), p. 557.

28. A. Friedman, One dimensional Stefan problems with nonmonotone free boundary, Trans. Amer. Math. Soc., vol. 133 (1968), pp. 89-114.

29. A. Friedman, Free boundary problems for parabolic equations, Bulletin Amer. Math. Soc., (1970), pp. 934-941.

30. A. Friedman, Analyticity of the free boundary for the Stefan problem, to appear.

31. A. Friedman and D. Kinderlehrer, A class of parabolic quasi-variational inequalities, J. of Diff. Equations, vol. 21(1976), pp. 395-416.

32. A. Friedman and D. Kinderlehrer, A one phase Stefan problem, Indiana U. Math. J., vol. 24(1975), pp. 1005-1035.

33. W. Fulks and R.B. Guenther, A free boundary problem and an extension of Muskat's model, Acta. Mathematica, vol. 122(1969), pp. 273-300.

34. D.B. Henry, Periodic solutions of the two-phase Stefan problem, to appear.

35. C.D. Hill and D.B. Kotlow, Classical solutions in the large of a two-phase free boundary problem, I, Arch. Rat. Mech. and Anal., vol. 45(1972), pp. 63-78.

36. C.D. Hill and D.B. Kotlow, Classical solution in the large of a two-phase free boundary problem, II, preprint.

37. A. Huber, Uber das Fortschreiten der Schmelzgrenze in einem linearen Leiter, AZMM, vol. 19(1939), pp. 1-21.

38. D. Kinderlehrer, The free boundary determined by the solution to a differential equation, Indiana Univ. Math. J., vol. 25(1976), pp. 195-208.

39. O.A. Ladyzenskaja, V.A. Solonnikov, and N.N. Ural'ceva, Linear and Quasilinear Equations of Parabolic Type, vol. 23, Translations of Math. Mono., Amer. Math. Soc., Providence, R.I., 1968.

40. J.L. Lions, Quelques Methodes de Resolution des Problemes uax Limites Non Lineaires, Dunod, Editeur, Gauthier-Villars, Paris, 1969.

41. M. Mori, Stability and convergence of a finite element method for solving the Stefan problem, Pub. of Res. Inst. for Math. Sci., Kyoto Univ., vol. 12(1976), pp. 539-563.

42. M. Muskat, Two fluid systems in porous media. The encorachment of water into an oil sand. Physics, vol. 5(1934), pp. 250-264.

43. L.I. Rubinstein, The Stefan Problem, vol. 27, Translations of Mathematical Monographs, American Math. Soc., Providence, R.I., 02904, 1967.

44. D.G. Schaefer, A new proof of the infinite differentiability of the free boundary in the Stefan problem, J. Diff. Eq., vol. 20(1976), pp. 226-269.

J. R. Cannon
Department of Mathematics
The University of Texas
RLM 8.100
Austin, Texas 78712
USA

MOVING BOUNDARY PROBLEMS

ONE PHASE MOVING BOUNDARY PROBLEMS

Avner Friedman

The purpose of this talk is to state some (mostly) recent results and indicate some problems in the area of one phase moving boundary problems.

1. EXISTENCE AND REGULARITY

Consider the Stefan problem: find functions $u(x,t)$, $s(t)$ satisfying:

$$u_{xx} - u_t = 0, \ 0 < x < s(t), \ 0 < t < T, \tag{1.1}$$

$$u(x,0) = h(x), \ 0 < x < b \ (b = s(0), h(b) = 0, \ h(x) > 0 \ \text{if} \ x \neq b) \tag{1.2}$$

$$u(0,t) = f(t), \ 0 < t < T \quad (f > 0) \tag{1.3}$$

$$u(s(t),t) = 0, \ 0 < t < T, \tag{1.4}$$

$$-u_x(s(t),t) = s'(t), \ 0 < t < T. \tag{1.5}$$

The curve $x = s(t)$ is called the moving boundary, or the free boundary.

It is well known that this problem has a unique solution in the usual (classical) sense. There are several methods to construct a solution: (1) reduce the problem to an integral equation for $u_x(s(t),t)$, (2) finite differences, (3) other approximation techniques, (4) weak solutions; see [3] [5] [10] [12] [13] [23] [27] and to the references given there.

The above problem arises as a model for melting of ice. But there are numerous other problems which give rise to free boundary problems, in which some of the conditions (1.3)-(1.5) become

ISBN 0-12-757350-X

different and also h may not remain positive; see [7] [8] [9] [17] [25] [27] [29].

Most of the existence proofs remain valid for a large class of free boundary problems provided T is sufficiently small. Thus in order to establish global existence in time, it suffices to show that we can carry out the local existence proof step-by-step on time intervals that do not shrink. This will be the case provided we can establish an a priori estimate

$$|u_x(x,t)| \le A \quad \text{if} \quad 0 < x < s(t),\ 0 < t < \sigma \tag{1.6}$$

whenever a solution u is known to exist for $t < \sigma$, where A is a positive constant independent of σ.

Such an a priori estimate has in fact been proved for (1.1)-(1.5) and for other free boundary problems as well; but not for general free boundary conditions. In fact we shall now give an example of non-existence in the sense that a solution $u(x,t)$, $s(t)$ exists for $t < T$, $s(T-0)$ is positive, but the solution "explodes" as $t \uparrow T$.

Melting of supercool water. If we drop the conditions $h > 0$, $f > 0$ then the temperature u may become negative, and we then conceive the water to be supercool in the regions where $u < 0$. Replace now (1.3) by

$$u_x(0,t) = 0, \quad t > 0 \tag{1.7}$$

and assume that

$$h(x) < 0 \quad \text{if} \quad 0 < x < b, \quad h(b) = 0. \tag{1.8}$$

Then $u(x,t) < 0$ if $0 \le x < s(t)$ and $s'(t) \le 0$ (by the maximum principle).

Let us make the further assumptions:

$$h'(0) = 0,\ h''(b) = (h'(b))^2,$$

$$h'(x) > 0,\ h''(x) > 0,\ 0 < x < b, \tag{1.9}$$

$$\frac{h''(x)}{h'(x)} \text{ is monotone decreasing,}$$

$$\int_0^b |h(x)|dx > b.$$

Theorem 1. Under the foregoing assumptions, there exists a $T < \infty$ such that the solution $u(x,t)$, $s(t)$ exists for $0 \le t < T$, and

$$\begin{aligned} &s'(t) \text{ is strictly decreasing for } 0 < t < T,\\ &s'(t) \to -\infty \text{ if } t \to T, \\ &s(T-0) > 0. \end{aligned} \tag{1.10}$$

Without assuming the first three conditions in (1.9) Sherman [31] proved that $s(T-0) > 0$, $\liminf s'(t) = -\infty$ as $t \to T$.

Theorem 1 is due to Friedman and Jensen [18]. Notice that the curve $x = s(t)$ is concave. Existence and regularity for melting of supercool (or mixed ordinary and supercool) water were established in [14] [22] [29]; in the first two papers the treatment employs methods of variational inequalities (see Section 4). The problem is related to optimal stopping time problems for a Brownian motion [29].

Among the various open problems for the existence of a solution, globally in time, let us mention the one in which (1.4), (1.5) are replaced by

$$u_x(s(t),t) = 0,\ 0 < t < T,$$

$$u(s(t),t) = 1 - \alpha \int_0^t \frac{d\tau}{1-s(\tau)},\ 0 < t < T;$$

it arises when a viscoplastic rod hits a rigid obstacle (see [25]); T is such that $s(T-0) = 0$. Existence for $0 < t < T_0$, T_0 sufficiently small, was proved in [30].

It is well known [4] [26] [28] that the free boundary for the problem (1.1)-(1.5) is C^∞. It was shown by Hill [21] that in general the free boundary is not an analytic function. The following is however true:

Theorem 2. If $f(t)$ is analytic for $0 < t < T$, then $s(t)$ is analytic for $0 < t < T$.

This is a result by Friedman [15] and it extends to free boundary problems where (1.4), (1.5) are replaced by more general conditions such as

$$u = \lambda_1, \quad u_x = \lambda_2 \dot{s} + \lambda_3 \qquad (\lambda_2 \neq 0)$$

where $\lambda_i = \lambda_i(s(t),t)$, $\lambda_i(x,t)$ analytic.

2. ASYMPTOTIC ESTIMATES

We consider the behavior of solutions as $t \to \infty$.

Theorem 3. Consider (1.1)-(1.5) (with $T = \infty$) and assume that $\int_0^\infty f(\tau)d\tau = \infty$, $f(t) \to 0$. Then

$$s(t) \sim \left(2 \int_0^t f(\tau)d\tau\right)^{1/2}.$$

If $f(t) \sim \alpha$ ($\alpha > 0$) then $s(t) \sim \beta t^{1/2}$ where β satisfies

$$\alpha = \sum_{n=1}^{\infty} \frac{n!}{(2n)!} \beta^{2n}.$$

This is proved in [32].

Consider next the free boundary problem:

$$u_{xx} - u_t = 0, \quad 0 < x < s(t), \quad t > 0 \tag{2.1}$$

$$u(x,0) = h(x), \quad 0 < x < b \quad (s(0) = b), \tag{2.2}$$

$$u_x(0,t) = -\ell(t), \quad t > 0 \quad (\ell(t) > -1), \tag{2.3}$$

$$u(s(t),t) = s(t), \quad t > 0, \tag{2.4}$$

$$u_x(s(t),t) = -s'(t), \quad t > 0. \tag{2.5}$$

This problem arises in a flow of water in a vertical porous pipe; u (the piezometric head) is equal to x + p, where p is the pressure of the water. Existence and uniqueness were established in [16].

To study the behavior, as $t \to \infty$, assume:

$$\beta \le h(0) + \int_0^t \ell(s)\,ds \le B \qquad (\beta > 0,\ B > 0) \tag{2.6}$$

(this ensures that s(t) does not shrink to zero in finite time and does not go to ∞ as $t \to \infty$),

$$\begin{aligned} &\int_0^\infty |\ell(t)|\,dt < \infty, \quad \sup_{0<t<\infty} |\ell'(t)| < \infty \\ &\ell(t) \to 0 \quad \text{if} \quad t \to \infty. \end{aligned} \tag{2.7}$$

Theorem 4. If (2.6), (2.7) hold, then

$$s(t) \to \gamma, \ s'(t) \to 0 \quad \text{as} \quad t \to \infty, \tag{2.8}$$

where

$$\gamma + \frac{1}{2}\gamma^2 = h(0) + b + \int_0^\infty \ell(t)\,dt. \tag{2.9}$$

This is proved in [17].

As a third example, consider the moving boundary problem:

$$u_{xx} + \frac{2}{x}u_x - u_t = 0, \ s(t) < x < \infty, \ t > 0, \tag{2.10}$$

$$\begin{aligned} &u(x,0) = h(x), \ x > b \ (b = s(0), \ h(b) = 0, \\ &\qquad h(x) > 0 \ \text{if} \ x \ne b) \end{aligned} \tag{2.11}$$

$$u(s(t),t) = s(t), \ t > 0, \tag{2.12}$$

$$\alpha u_x(s(t),t) = s(t)s'(t) + \alpha, \ t > 0 \ (\alpha \ne 0). \tag{2.13}$$

Here u represents the density of vapor surrounding a drop of radius $s(t)$; $\alpha < o$ corresponds to the case of condensation and $\alpha > 0$ corresponds to the case of evaporation. Existence and uniqueness were established in [8].

The parameter α is physically very small. It is therefore interesting to study the behavior of $s(t) = s(t,\alpha)$ $(\alpha < 0)$ as $t \to \infty$, $\alpha \to 0$ jointly. The following result is proved in [8]:

Theorem 5. If $\alpha < 0$, then

$$\overline{\lim_{t\to\infty}}\left|\frac{s^2(t)}{2\alpha t} - 1\right| \le \eta(\alpha),$$

$$\overline{\lim_{t\to\infty}}\left|\frac{s(t)s'(t)}{\alpha} + 1\right| \le \eta(\alpha),$$

where $\eta(\alpha) \to 0$ if $\alpha \to 0$.

If $\alpha = 0$ then we have an equilibrium and $s(t) \equiv b$. One can develop $s(t,\alpha)$ (for $0 \le t \le T$, $T < \infty$) into a power sereis in α:

$$s(t,\alpha) = b + \sum_{n=1}^{\infty} s_n(t)\alpha^n \tag{2.14}$$

(see [8]).

Results similar to Theorem 5 and to (2.14) have been obtained for the problem of a dissolution of a gas bubble in liquid [9]. It would be interesting to establish such results for other problems which arise in applications.

3. THE SHAPE OF THE FREE BOUNDARY

For the solution of the Stefan problem (1.1)-(1.5), $s(t)$ is strictly monotone increasing. What else can be said about its shape?

Let us assume that

$$f(t) \equiv f(0) = h(0),$$

$$h''(0) = 0,\ h''(b) = (h'(b))^2,$$

$$h'(x) < 0,\ h''(x) > 0,\ 0 < x < b, \tag{3.1}$$

$$\frac{h''(x)}{h'(x)} \text{ is monotone increasing.}$$

Theorem 6. If (3.1) holds then $s'(t)$ is strictly monotone decreasing and $\lim_{t\to\infty} s'(t) = 0$.

Thus $x = s(t)$ is a concave curve.

This is a recent result of Friedman and Jensen [18]. If the last condition in (3.1) is replaced by:

$$\frac{h''(x)}{h'(x)} \text{ is piecewise monotone,}$$

then the same is true of $s'(t)$; in fact, $s'(t)$ changes the direction of monotonicity at most as many times as it does the function h''/h'.

Problem. Establish convexity-type results for other moving boundary problems.

For some problems, $s(t)$ is not necessarily monotone, but we can still determine an upper bound (in terms of the data) on the number of times that $s'(t)$ changes sign. We give an example from the hydraulic problem (2.1)-(2.5).

Theorem 7. [17]. Assume that $h'(x)$ changes sign H times and $\ell(t)$ changes sign L times. Then $s'(t)$ changes sign at most $H + L$ times.

A result of this type is valid also for the Stefan problem with supercool water [14] [29]: If $h(x)$ changes sign at most H times then $s'(t)$ changes sign at most H times.

It would be interesting to establish such results for other problems.

4. CONNECTION WITH VARIATIONAL INEQUALITIES

Consider the problem: Find a function v satisfying:

$$v_t - v_{xx} \geq f, \; v \geq \phi, \; (v_t - v_{xx} - f)(v - \phi) = 0 \text{ a.e.} \quad \text{for } x > 0, \; t > 0, \tag{4.1}$$

$$v(x,0) = k(x), \; x > 0 \quad (k \geq \phi),$$

$$v(0,t) = \ell(t), \; t > 0 \quad (\ell \geq \phi).$$

This problem is called a variational inequality. One can solve it by standard methods for parabolic equations, involving well known a priori L^p-estimates. The solution u is generally smooth in the sense that the distribution derivatives

$$u_t, u_x, u_{xx} \text{ are in } L^p;$$

u_t and u_{xx} are not continuous in general.

Suppose now that

$$v(x,t) > \phi(x,t) \text{ if and only if } 0 < x < s(t)$$

and set

$$u = \frac{\partial}{\partial t}(v - \phi). \tag{4.4}$$

Then u satisfies

$$u_t - u_{xx} = F, \; 0 < x < s(t), \; t > 0, \tag{4.5}$$

$$u(x,0) = h(x), \; 0 < x < b \quad (b = s(0)), \tag{4.6}$$

$$u(0,t) = f(t), \; t > 0, \tag{4.7}$$

$$u(s(t),t) = 0, \; t > 0, \tag{4.8}$$

$$u_x(s(t),t) = \lambda(s(t),t)s'(t), \quad t > 0 \tag{4.9}$$

where

$$F = f-(\phi_t-\phi_{xx}), \; h = k'' - \phi_t, \; f = (\ell-\phi)_t, \; \lambda = \phi_{xx} - \phi_t.$$

Thus, the variational inequality reduces to a moving boundary problem; if $F = 0$, $\lambda = -1$, $h > 0$, $f > 0$, this is precisely the Stefan problem (1.1)-(1.5).

Conversely, one can go from a free boundary problem into a variational inequality by reversing the transformation (4.4):

$$v(x,t) = \int_{\sigma(x)}^{t} u(x,\eta)d\eta + \phi(x,t) \tag{4.10}$$

where σ is the inverse of s; this requires that the free boundary $x = s(t)$ be monotone or, in some cases, just piecewise monotone. It enables us to study free boundary problems by the methods of variational inequalities (or vice versa); [14] [22] [29].

Actually, if the conditions on the free boundary are not those occurring in the Stefan problem, we do not get in general a variational inequality of the form (4.1)-(4.3). Thus, in the case of

$$u(s(t),t) = 0, \quad t > 0, \tag{4.11}$$

$$u_x(s(t),t) = -\lambda_x(s(t),t)s'(t)-\lambda_t(s(t),t), \quad t > 0$$

we get the problem (4.1)-(4.3) for v defined by (4.10), with f which is a function of x,t and $s(t)$. Such a problem is called a <u>quasi variational inequality</u>; it was studied in [20] [16].

A general open problem is to study free boundary problems by reducing them, in some manner, to quasi variational inequalities.

5. THE n-DIMENSIONAL STEFAN PROBLEM

We shall give several formulations. Let G be a bounded domain in R^n whose boundary consists of two smooth surfaces: the interior boundary Γ_i and the exterior boundary Γ_e. Denote by G_i the domain bounded by Γ_i and set $D = B \setminus G_i$ where B is a large ball with center 0 containing G.

Problem 1. Find functions $\theta(x,t)$, $t = s(x)$ $(x \in D,$ $0 < t < T)$ satisfying:

$$s(x) = 0 \text{ if } x \in G_i \tag{5.1}$$

$$-\Delta\theta + \theta_t = 0 \text{ if } x \in D,\ s(x) < t, \tag{5.2}$$

$$\theta = 0 \text{ if } t = s(x),\ x \in D\setminus G, \tag{5.3}$$

$$\nabla\theta\cdot\nabla s = -k \text{ if } t = s(x),\ x \in D\setminus G, \tag{5.4}$$

$$\theta = h \text{ if } x \in G,\ t = 0 \tag{5.5}$$

$$\theta = g \text{ if } x \in \Gamma_i,\ 0 < t < T. \tag{5.6}$$

Here $k > 0$ is a constant and $g(x,t)$, $h(x)$ are positive and smooth.

Problem 1 is the classical formulation of the Stefan problem. The trouble with it is that a solution may not exist in general. We therefore proceed to another formulation involving weak solutions:

Problem 2. Find a bounded measurable function $\theta(x,t)$ such that

$$\int_0^T \int_D [\theta\Delta\zeta + a(\theta)\zeta_t]dxdt = \int_0^T \int_{\partial D} g\frac{\partial\zeta}{\partial\nu}\, dSdt - \int_D a(\theta_0)\zeta(x,0)dx, \tag{5.7}$$

where ν is the outward normal, for any function ζ which is smooth in $\overline{D} \times [0,T]$, $\zeta = 0$ if either $x \in \partial D$ or $t = T$, and

$$a(\theta) = \begin{cases} \theta & \text{if } \theta > 0, \\ \theta - k & \text{if } \theta \leq 0, \end{cases}$$

$$a(\theta_0) = \begin{cases} h(x) & \text{if } x \in G, \\ -k & \text{if } x \in D \setminus G. \end{cases}$$

In this formulation, there exists a unique solution [12] [23].

We proceed with a variational inequality formulation (this is due to Duvaut [6]) for the function u defined by

$$u(x,t) = \int_{s(x)}^{t} \theta(x,\tau)d\tau \text{ if } x \in D\setminus G,\ s(x) \leq t < T,$$

$$u(x,t) = 0 \text{ if } x \in D\setminus G,\ 0 \leq t < s(x), \tag{5.8}$$

$$u(x,t) = \int_0^t \theta(x,\tau)d\tau \text{ if } x \in G,\ 0 \leq t < T.$$

Problem 3. Find a function u satisfying:

$$-\Delta u + u_t \geq f,\ u \geq 0,\ (-\Delta u + u_t - f)u \geq 0$$

$$\text{a.e. if } x \in D,\ 0 < t < T; \tag{5.9}$$

$$u = \psi \text{ on } \Gamma_i \times (0,T), \tag{5.10}$$

$$u = 0 \text{ on } \partial B \times (0,T), \tag{5.11}$$

$$u = 0 \text{ on } D \times \{0\}, \tag{5.12}$$

where

$$\psi(x,t) = \int_0^t g(x,\tau)d\tau,$$

$$f(x) = \begin{cases} h(x) & \text{if } x \in G, \\ -k & \text{if } x \in D \setminus G. \end{cases}$$

It can be shown [19] that θ solves Problem 2 if and only if u solves Problem 3.

The formulation (5.9)-(5.12) enables one to use more effectively the general theory of partial differential equations. We shall state some recent results for the solution u of Problem 3.

Assume first that

$$\Gamma_i \text{ is a sphere with center } 0. \tag{5.13}$$

Writing in polar coordinates $\Delta w = \rho^{1-n}(\rho^{n-1}w_\rho)_\rho + \rho^{-2}Aw$, we also assume that

$$\psi_t(x,t)-\psi_t(x,0)-\rho^{-2}A\psi(x,t) \geq 0 \text{ on } \Gamma_i \times (0,T), \tag{5.14}$$

$$h(x) > 0 \text{ in } G,\ h(x) = 0 \text{ on } \Gamma_e, \tag{5.15}$$

$$(\rho^2 h)_\rho < 0 \text{ in } G$$

and

$$h(x) = \psi_t(x,0) = g(x,0) \text{ for } x \in \Gamma_i. \tag{5.16}$$

Theorem 8. Under the assumptions (5.13)-(5.16), there exists a function $\rho = \rho^*(\theta,t)$ such that

$$u(x,t) > 0 \text{ if and only if } \rho < \rho^*(\theta,t).$$

Further, $\rho^*(\theta,t)$ is Lipschitz continuous in θ and continuous and monotone decreasing in t.

This result, which is due to Friedman and Kinderlehrer [19] extends also to Γ which is "nearly" spherical.

Denote the set occupied by ice by I. Suppose (x_0,t_0) is a point on the boundary of both the set I and the set occupied by water, i.e., (x_0,t_0) belongs to the moving (or free) boundary. Suppose also that $0 < t_0 < T$ and

$$\text{the set } I \cap \{t = t_0\} \text{ has positive density at } (x_0,t_0). \tag{5.17}$$

Theorem 9. If u is the solution of Problem 3 and if (5.17) holds, then there exists a neighborhood N of (x_0,t_0) such that the free boundary in N is a C^2 surface S and all the second derivatives of u are continuous up to S.

This is a deep result due to L. Caffarelli [1]. He also studied in [2] the set of points where the condition (5.17) does not hold and showed that it cannot evolve smoothly in time.

Using a hodograph type mapping Kinderlehrer and Nirenberg [24] showed that the assertion of Theorem 9 can be strengthened: the free boundary S (in N) is C^∞ and u is C^∞ up to the boundary S. Combining this with Theorem 8 we see that, under the conditions of Theorem 8, the free boundary is everywhere C^∞ and the solution is C^∞ up to the boundary.

The general theory of variational inequalities implies that the distribution derivatives u_t, u_{x_i}, $u_{x_i x_i}$ are in L^∞; for the special case of the Stefan problem one can show [19] that also u_{tx_i} and u_{tt} are in L^2. But the continuity of $u(x,t)$ is still an open question.

A partial result was recently given by Caffarelli [2]:

Theorem 10. If $n = 2$ then $u(x,t)$ is continuous.

REFERENCES

1. L. A. Caffarelli, The regularity of the free boundaries in higher dimensions, Acta Math., to appear.
2. L. A. Caffarelli, Some aspects of the one-phase Stefan problem, Indiana Univ. Math. J., to appear.
3. J. R. Cannon and C. D. Hill, Existence, uniqueness, stability, and monotone dependence in a Stefan problem with the heat equation, J. Math. Mech., 17 (1967), 1-19.
4. J. R. Cannon and C. D. Hill, On the infinite differentiability of the free boundary in the Stafan problem, J. Math. Anal. Appl., 22 (1968), 385-397.
5. A. Datzeff, Sur le problème linéaire de Stefan, Gauthier-Villars, Paris, 1970.
6. G. Duvaut, Résolution d'un problème de Stefan (Fusion d'un bloc de glace à zero degré), C. R. Acad. Sc. Paris, 276 (1973), 1461-1463.
7. A. Fasano and M. Primicerio, General free boundary problems for the heat equation, Part I, J. Math. Anal. Appl., 57 (1977), 694-723; Part II, ibid. 58 (1977), 202-231; Part III, ibid, 59 (1977), 1-14.
8. A. Friedman, Free boundary problems for parabolic equations II. Evaporation or condensation of a liquid drop, J. Math. Mech., 9 (1960), 19-66.
9. A. Friedman, Free boundary problems for parabolic equations III. Dissolution of a gas bubble in liquid, J. Math. Mech., 9 (1960), 327-345.
10. A. Friedman, Partial Differential Equations of Parabolic Type, Prentice-Hall, Englewood Cliffs, N. J., 1964.
11. A. Friedman, Asymptotic behavior of solutions of parabolic differential equations and of integral equations, Differential Equations and Dynamical Systems, Academic Press, New York, 409-426, 1967.
12. A. Friedman, The Stefan problem in several space variables,

Trans. Amer. Math. Soc., 133 (1968), 51-87.

13. A. Friedman, One dimensional Stefan problems with non-monotone free boundary, Trans. Amer. Math. Soc., 133 (1968), 89-114.
14. A. Friedman, Parabolic variational inequalities in one-space dimension and smoothness of the free boundary, J. Funct. Analys., 18 (1975), 151-176.
15. A. Friedman, Analyticity of the free boundary for the Stefan problem, Archive Rat. Mech. Anal., 61 (1976), 97-125.
16. A. Friedman, A class of parabolic quasi-variational inequalities II, J. Diff. Eqs., 22 (1976), 379-401.
17. A. Friedman and R. Jensen, A parabolic quasi-variational inequality arising in hydraulics, Ann. Scu. Norm. Sup. Pisa, 2 (4) (1975), 421-468.
18. A. Friedman and R. Jensen, Convexity of the free boundary in the Stefan problem and in the dam problem, Archive Rat. Mech. Anal., to appear.
19. A. Friedman and D. Kinderlehrer, A one phase Stefan problem, Indiana Univ. Math. J., 24 (1975), 1005-1035.
20. A. Friedman and D. Kinderlehrer, A class of parabolic quasi-variational inequalities, J. Diff. Eqs., 21 (1976), 395-416.
21. C. D. Hill, Parabolic equations in one space variable and the non-characteristic Cauchy problem, Comm. Pure Appl. Math., 20 (1967), 619-633.
22. R. Jensen, Smoothness of the free boundary in the Stefan problem with supercooled water, Ill. J. Math., to appear.
23. S. L. Kamenomostkaja, On Stefan's problem, Math. Sbornik, 53 (95) (1965), 485-514.
24. D. Kinderlehrer and L. Nirenberg, Regularity in the free boundary problems, Ann. Scu. Norm. Sup. Pisa, to appear.
25. S. N. Kruzkov, On some problems with unknown boundaries for the heat conduction equation, J. Appl. Math. Mech., 31 (1967), 1014-1024.
26. C. Li-Shang, Existence and differentiability of the solution of the two-phase Stefan problem for quasi-linear parabolic

equations, Chinese Math. -Acta, 7 (1965), 481-496.

27. L. I. Rubinstein, The Stefan Problem, Translations of Mathematical Monographs, vol. 27, Amer. Math. Soc., Providence, R.I., 1971.
28. D. G. Schaeffer, A new proof of the infinite differentiability of the free boundary in the Stefan problem, J. Diff. Eqs., 20 (1976), 266-269.
29. P. Van Moerbeke, An optimal problem for linear reward, Acta Math., 132 (1974), 1-41.
30. A. Fasano and M. Primicerio, Viscoplastic impact of a rod on a wall, Bull. Union Mat. Ital., (4) 11 (1975), 531-553.
31. B. Sherman, A general one-phase Stefan problem, Quar. Appl. Math., 28 (1970), 377-382.
32. J. R. Cannon and C. D. Hill, Remarks on a Stefan problem, J. Math. Mech., 17 (1967), 433-440.

Northwestern University
Department of Mathematics
Evanston, Illinois 60201

This work is partially supported by National Science Foundation Grant MC575-21416 A01.

MOVING BOUNDARY PROBLEMS

APPLICATION OF CLASSICAL ANALYTICAL TECHNIQUES TO STEADY-STATE FREE BOUNDARY PROBLEMS

Bernard A. Fleishman, Ross Gingrich, Thomas J. Mahar

Elliptic partial differential equations, of the form div(K grad u) + f = 0, are considered, in which K and/or f depends discontinuously on the dependent variable u. Discontinuous nonlinearities of this type give rise to free boundaries (moving boundaries in time-dependent problems). Such a PDE describes, for example, a steady-state Stefan problem. We present here two examples of application of a hybrid perturbation (or linearization) method to special cases of the above PDE, to obtain approximate solutions of two-dimensional free boundary problems.

1. INTRODUCTION

Linearization methods have not been used much to attack free or moving boundary problems, no doubt because any such problem involves a discontinuity in an essential way. Yet such methods, if they work, can be practical and easy to use.

For the past two years the authors have been engaged in investigating the applicability of classical approximation techniques to free boundary problems; specifically, to certain boundary value problems for second order nonlinear elliptic PDE's in divergence form,

$$\operatorname{div}(K \operatorname{grad} u) + f = 0 \quad (K > 0) \tag{1.1}$$

in which K and/or f may be a piecewise-continuous function of u (and perhaps also of the independent variables). (See [1,2,3] for a variety of results for the one-dimensional version of (1.1).)

ISBN 0-12-757350-X

We present here two examples of application of a hybrid perturbation (more precisely, linearization) method to special cases of equation (1.1), to obtain approximate solutions of two-dimensional free boundary problems. The first to be considered, a steady-state Stefan problem, (see [7]) is treated in Sections 2-5. The second example, in Section 6, is discussed more briefly for it has been given a detailed treatment elsewhere [5].

Both problems contain a small parameter ε; when $\varepsilon = 0$ the reduced problem, one-dimensional but containing the (discontinuous) nonlinearity, has a solution u_o. When we seek a perturbed solution of the form $u_o(x)+\varepsilon\tilde{u}(x,y)$, by linearizing about u_o, the boundary value problem (BVP) for $\tilde{u}$, though linear, is found to have interesting features, somewhat different in each problem. These arise from the fact that our variational equation, the linear PDE to be satisfied by $\tilde{u}$, results from differentiating a PDE containing a step-function of u.

Our procedure is a formal one; we have not yet obtained error bounds. But as noted in Section 5, the Stefan problem can be solved exactly by another method, and the results of the two methods are in excellent agreement. There are also indications, mentioned in Section 6, of the reasonableness of our results in the other problem. In particular, the existence of a solution to a similar BVP has been proved [5,6] by an iterative method.

2. SSSP (STEADY STATE STEFAN PROBLEM): FORMULATION

In the strip

$$S = \{(x,y):\ 0 < x < 1,\ -\infty < y < \infty\}$$

of the x,y-plane we consider the boundary value problem, which will be called $P_1(\varepsilon)$:

$$\frac{\partial}{\partial x}\left(K \frac{\partial u}{\partial x}\right) + \frac{\partial}{\partial y}\left(K \frac{\partial u}{\partial y}\right) = 0 \qquad \text{in } S \tag{2.1}$$

$$u(0,y) = T_1+\varepsilon h(y),\ u(1,y) = T_2,\ -\infty < y < \infty \tag{2.2}$$

where T_1 and T_2 are constants, $\varepsilon \geq 0$ is a small parameter, h is a continuous, periodic function, and $K = K(u)$ is the step-function

$$K(u) = \begin{cases} k_1 & \text{for } u < \mu \\ k_2 & \text{for } u \geq \mu \end{cases} \tag{2.3}$$

for constant μ and positive constants k_1 and k_2. Then the constants T_1, T_2, ε, μ are assumed to satisfy (for $|h(y)| \leq 1$)

$$T_1 + \varepsilon < \mu < T_2. \tag{2.4}$$

A standard interpretation of equation (2.1) is that it governs time-independent heat conduction (in two dimensions), u denoting the temperature and K the heat conductivity of the medium. The jump in K when u passes through the value μ models the fact that conductivity can change discontinuously when the medium experiences a change of state.

In this context, assumption (2.4), relating the boundary temperatures to the threshold value μ, tells us that near the boundary $x = 0$ the medium is in one phase (say, solid) while near the other boundary it is in another phase (say, liquid). Thus, if u is continuous there must be one or more curves where $u = \mu$. To determine such free boundaries, (so called, because they are not known a priori) is a point of primary interest in the solution of $P_1(\varepsilon)$.

We seek a solution of $P_1(\varepsilon)$ such that $u(x,y)$ is continuous and bounded in $\bar{S}$ and C^2 at all points of S where $K(u(x,y))$ is continuous. Further, across a free boundary we require the "flux" $K(u)\partial u/\partial n$ (where $\partial u/\partial n$ is the normal derivative) to be continuous.

3. SSSP: THE REDUCED PROBLEM $P_1(0)$

When $\varepsilon = 0$ the boundary conditions (2.2) are both independent of y, and $P_1(0)$ reduces to a one-dimensional BVP:

$$P_1(0)\begin{cases}(K(u)u')' = 0 \text{ in } I: 0 < x < 1 & (3.1)\\ u(0) = T_1,\ u(1) = T_2 & (3.2)\end{cases}$$

where $' = d/dx$.

Let u_o denote a solution of $P_1(0)$ with the required regularity: $u_o(x)$ and $K(u_o(x))u_o'(x)$ are continuous in I. Equation (3.1) implies that $K(u_o(x))u_o'(x)$ is also piecewise-constant in I. Hence

$$K(u_o(x))u_o'(x) = \phi \text{ (constant) in } I.$$

Since $K > 0$ always, u_o' has the same sign as ϕ. $\phi \leq 0$ would imply that u_o is non-increasing in I, which is impossible since $T_1 < \mu < T_2$. Thus $\phi > 0$, and $u_o' > 0$ (wherever it is defined).

It follows that u_o is strictly increasing in I, from T_1 to T_2. Since u_o is also continuous, there is a unique value of x in I, x_o, such that

$$u_o(x_o) = \mu,$$

and $K(u_o(x))$ and $u_o'(x)$ are discontinuous only at $x = x_o$. On I_1: $0 < x < x_o$, $u_o(x) < \mu$ and $K(u_o(x)) = k_1$, while on

I_2: $x_o < x < 1$, $u_o(x) > \mu$ and $K(u_o(x)) = k_2$.

Integrating $u_o'(x) = \phi/k_i$ on I_i $(i = 1,2)$, then imposing boundary conditions (3.2) plus the continuity and interface conditions $u_o(x_o-) = u_o(x_o+) = \mu$, gives

$$u_o(x) = \begin{cases} \phi x/k_1 + T_1, & 0 \leq x \leq x_o \quad (3.3) \\ \phi(x-1)/k_2 + T_2, & x_o \leq x \leq 1 \quad (3.4) \end{cases}$$

where

$$x_o = k_1(\mu - T_1)/\phi, \qquad \phi = k_1(\mu - T_1) + k_2(T_2 - \mu) . \qquad (3.5)$$

4. SSSP: LINEARIZATION

We seek an (approximate) solution of $P_1(\varepsilon)$ in the form

$$u(x,y) \simeq u_o(x) + \varepsilon\tilde{u}(x,y) \qquad (4.1)$$

where $\tilde{u}$ is periodic in y and uniformly bounded in $\bar{S}$. Similarly we assume that the solution (4.1) has a free boundary which may be represented

$$x \simeq x_o + \varepsilon g(y) , \qquad (4.2)$$

that is, as a perturbation of the free boundary $x = x_o$ in the reduced problem. (In (4.1) and (4.2) and in the manipulations to follow, terms of order ε^2 are neglected.)

Expanding $K(u) = K(u_o + \varepsilon\tilde{u})$ formally about $u = u_o$, we have

$$K(u) \simeq K(u_o) + K'(u_o)\varepsilon\tilde{u} . \qquad (4.3)$$

Upon substitution for u and K(u) in (2.1) by use of (4.1) and

(4.3), the PDE takes the form

$$[(K(u_o) + \varepsilon K'(u_o)\tilde{u})(u_o+\varepsilon\tilde{u})_x]_x$$

$$+[(K(u_o) + \varepsilon K'(u_o)\tilde{u})(u_o+\varepsilon\tilde{u})_y]_y = 0$$

or, neglecting terms which are $O(\varepsilon^2)$ and noting that u_o and $K(u_o)$ are functions only of x,

$$[K(u_o)u_o' + \varepsilon\{K'(u_o)u_o'\tilde{u} + K(u_o)\tilde{u}_x\}]_x$$

$$+ \varepsilon K(u_o)\tilde{u}_{yy} = 0$$

or, since $(K(u_o)u_o')' = 0$,

$$(K(u_o)\tilde{u})_{xx} + (K(u_o)\tilde{u})_{yy} = 0 \quad \text{in } S. \tag{4.4}$$

The boundary conditions for $\tilde{u}$ follow from (4.1), $u_o(0) = T_1$, $u_o(1) = T_2$ and (2.1):

$$\tilde{u}(0,y) = h(y),\ \tilde{u}(1,y) = 0,\ -\infty < y < \infty . \tag{4.5}$$

We can solve the (linear) BVP (4.4), (4.5) for $\tilde{u}$ by separation of variables; therefore it suffices to carry out the details in the case

$$h(y) = \cos y.$$

More general periodic functions h may be handled by Fourier decomposition (e.g., see [5], pp. 566-7).

Substitution of

$$\tilde{u}(x,y) = v(x)\cos y$$

in (4.4) and (4.5) (with $h(y) = \cos y$) yields the BVP

$$(K(u_o)v)'' - K(u_o)v = 0 \quad \text{in } I \tag{4.6}$$

$$v(0) = 1,\ v(1) = 0\ . \tag{4.7}$$

The question arises now as to how much regularity we should require of the solution v of this BVP. If we ask that v be continuous on I, $K(u_o)v$ will have a jump discontinuity at $x = x_o$ (where $u(x_o) = \mu$) while the first term in (4.6) experiences a singularity of higher order. For this as well as other reasons, we reject the requirement that v be continuous on I.

Rather, guided by the form of equation (4.6), we set

$$w(x) = K(u_o(x))v(x)$$

in (4.6) and (4.7) and obtain the BVP

$$w'' - w = 0 \qquad \text{in } I \tag{4.8}$$

$$w(0) = k_1,\ w(1) = 0. \tag{4.9}$$

(Recall that $K(u_o(0)) = k_1$.) Now in view of (4.8) we may as well seek a solution w which is not only continuous on I, but C_2! An elementary calculation yields

$$w(x) = \frac{k_1}{\sinh 1} \sinh(1-x) \tag{4.10}$$

on I_1. Then from $w(x) = K(u_o(x))v(x)$ and the fact that $K(u_o(x)) = k_i$ on I_i $(i = 1,2)$, we find

$$v(x) = \begin{cases} \dfrac{\sinh(1-x)}{\sinh 1}, & 0 \leq x \leq x_o \quad (4.11) \\[2ex] \dfrac{k_1}{k_2}\dfrac{\sinh(1-x)}{\sinh 1}, & x_o \leq x \leq 1 \quad (4.12) \end{cases}$$

That v has a discontinuity at $x = x_o$ may be understood as follows. The PDE to be satisfied by $\tilde{u}(x,y) = v(x)\cos y$, equation (4.4), is the variational equation for (2.1), derived, in effect, from (2.1) by differentiation (see (4.3)). It is not surprising that in the process we lose one order of regularity, that while u is continuous, with piecewise-continuous derivatives, $\tilde{u}$ (and therefore v) is piecewise-continuous.

The handle we have for adjusting the free boundary to take account of the perturbation in the boundary conditions, is the basic requirement that $u = \mu$ along some curve of the form (4.2). Substitution of $x_o + \varepsilon g(y)$ for x in

$$u(x,y) = u_o + \varepsilon\tilde{u}(x,y) = u_o(x) + \varepsilon v(x)\cos y = \mu$$

gives

$$u_o(x_o+g(y)) + \varepsilon v(x_o+\varepsilon g(y))\cos y = \mu$$

$$\frac{\phi}{k_1}(x_o+\varepsilon g(y)) + T_1 + \frac{\varepsilon \cos y}{\sinh 1}\sinh(1-x_o-\varepsilon g(y)) = \mu \qquad (4.13)$$

where we have used the representations (3.3) and (4.11) for u_o and v, respectively. Noting that $u_o(x_o) = \phi x_o/k_1 + T_1 = \mu$ and dropping terms of order ε^2, we obtain

$$\varepsilon(\phi/k_1)g(y) + \varepsilon \cos y \sinh(1-x_o)/\sinh 1 = 0$$

and finally

$$g(y) = -\frac{k_1}{\phi}\frac{\sinh(1-x_o)}{\sinh 1}\cos y. \tag{4.14}$$

To sum up, we have derived an approximate solution of $P_1(\varepsilon)$ (with $h(y) = \cos y$, K given by (2.3), and the constants T_1, T_2, μ and the small parameter $\varepsilon \geq 0$ satisfying (2.4)) in the form

$$u(x,y)=\begin{cases} \dfrac{\phi x}{k_1} + T_1 + \varepsilon\dfrac{w(x)}{k_1}\cos y, & 0 \leq x \leq x_o + \varepsilon g(y) \\[2ex] \dfrac{\phi(x-1)}{k_2} + T_2 + \varepsilon\dfrac{w(x)}{k_2}\cos y, & x_o+\varepsilon g(y) \leq x \leq 1 \end{cases} \tag{4.15}$$

where the constants x_o and ϕ are given by (3.5), while the functions w and g are given by (4.10) and (4.14) respectively.

5. SSSP: CRITIQUE

First we remark that there is internal consistency in our results. It is easily seen that up through terms of first order in ε the function u, given by (4.15), is continuous across the interface $x = x_o + \varepsilon g(y)$, g given by (4.14), and so is the flux $K\,\partial u/\partial n$. Also, the boundary conditions (2.2) are satisfied, and the expressions for u in (4.15) satisfy, in their respective domains, the PDE (2.1). (Since K has one constant value, k_1, to the left of the interface and another, k_2, to the right, in each region (2.1) reduces to Laplace's equation.)

Secondly, we have a check on the validity of our results. As noted in Section 1, under the Kirchhoff transformation

$$U(x,y) = \int_0^{u(x,y)} K(s)\,ds$$

problem $P_1(\varepsilon)$ reduces to a classical Dirichlet problem for Laplace's equation. This problem for U may be solved exactly, and when the results are interpreted for u they agree, up through terms of first order in ε, with the results for u and g, given in (4.15) and (4.14), obtained by our formal approximation technique.

6. A SECOND FREE BOUNDARY PROBLEM

Consider the BVP $P_2(\varepsilon)$ defined by

$$\partial^2 u/\partial x^2 + \partial^2 u/\partial y^2 + f(u) = 0 \qquad \text{in } S \tag{6.1}$$

$$u(0,y) = \varepsilon \cos y, \; u_x(1,y) = 0, \; -\infty < y < \infty \tag{6.2}$$

where $\varepsilon \geq 0$ is a small parameter and f is a step-function with threshold value $\mu > 0$:

$$f(u) = \begin{cases} 0 & \text{for } u < \mu \\ 1 & \text{for } u \geq \mu \end{cases} \tag{6.3}$$

We shall assume that

$$0 \leq \varepsilon < \mu.$$

Since the treatment of $P_2(\varepsilon)$ by our approximation technique has been derived elsewhere [5] in detail, here the procedure will be described only briefly.

Again we seek an approximate solution of $P_2(\varepsilon)$ in the form

$$u(x,y) = u_o(x) + \varepsilon\tilde{u}(x,y), \tag{6.4}$$

where u_o is a solution of the reduced problem

$$P_2(0)\begin{cases} u'' + f(u) = 0 \quad \text{in I:} \quad 0 < x < 1 \\ u(0) = 0,\ u'(1) = 0. \end{cases}$$

For $\mu > 0$, $P_2(0)$ always has the trivial solution $u(x) \equiv 0$. In addition, when $0 < \mu < 1/4$, it may be seen that for each of the roots $x_o (0 < x_o < 1)$ of the algebraic equation

$$x_o(1-x_o) = \mu \tag{6.5}$$

$P_2(0)$ has a non-trivial solution

$$u_o(x) = \begin{cases} (1-x_o)x, & 0 \leq x \leq x_o \\ x - \frac{1}{2}(x^2 + x_o^2), & x_o \leq x \leq 1. \end{cases} \tag{6.6}$$

Thus, for $0 < \mu < 1/4$, $P_2(0)$ has three solutions, the trivial one $u(x) \equiv 0$ (with respect to which $P_2(0)$ is a linear problem) and two non-trivial solutions $u_o(x)$ given by (6.6), each corresponding to one of the roots x_o of (6.5). With respect to the solutions u_o, $P_2(0)$ is a free boundary (that is, a nonlinear) problem, for $f(u_o(x))$ jumps in value at $x_o \in I$, and x_o is not known a priori.

For fixed $\mu \in (0,1/4)$ let u_o be the solution of $P_2(0)$ corresponding to the smaller root of equation (6.5); thus, $0 < x_o < 1/2$. On the basis of a rigorous existence proof for a similar problem, but in a circular domain, (see [5,6]) there is good reason to expect that $P_2(\varepsilon)$ possesses a y-periodic solution u close to this u_o, and we assume that u can be written, neglecting terms which are $O(\varepsilon^2)$, in the form (6.4), where $\tilde{u}$ is periodic in y and uniformly bounded in $\bar{S}$.

To derive a BVP for $\tilde{u}$, we first subtract $\Delta u_o + f(u_o) = 0$

from $\Delta u + f(u) = 0$, or $\Delta(u_o+\varepsilon\tilde{u}) + f(u_o+\varepsilon\tilde{u}) = 0$. Then, noting that (formally)

$$f(u_o + \varepsilon\tilde{u}) \simeq f(u_o) + \varepsilon\tilde{u}f'(u_o)$$

while $f'(u_o(x)) = \delta[u_o(x)-\mu] = \delta(x-x_o)/u_o'(x)$,
we obtain

$$\Delta\tilde{u} + \frac{\delta(x-x_o)}{u_o'(x)}\tilde{u} = 0 \quad \text{in } S. \tag{6.7}$$

From (6.2), (6.4), $u_o(0) = 0$ and $u_o'(1) = 0$, there follows

$$\tilde{u}(0,y) = \cos y,\ \tilde{u}_x(1,y) = 0,\ -\infty < y < \infty \tag{6.8}$$

Substitution in (6.7) and (6.8) of

$$\tilde{u}(x,y) = v(x) \cos y$$

yields the BVP

$$\begin{aligned} &v'' - v + [\delta(x-x_o)/u_o'(x)]v = 0 \quad \text{in } I \\ &v(0) = 1,\ v'(1) = 0. \end{aligned} \tag{6.9}$$

For v a continuous solution of the BVP (6.9) on $[0,1]$, the delta-function requires that v' undergo a jump at $x = x_o$:

$$v'(x_o+)-v'(x_o-) = -v(x_o)/u_o'(x_o) = -v(x_o)/(1-x_o) \tag{6.10}$$

We now solve $v'' - v = 0$ in each of the intervals $0 < x < x_o$ and $x_o < x < 1$. The four arbitrary constants are determined by the jump condition (6.10), the continuity condition $v(x_o+) = v(x_o-)$, and the boundary conditions in (6.9), and we obtain where x_0 is the smaller root of (6.5), $\bar{x}_0 = 1 - x_0$, A and B are given by (6.11), and g is given by (6.12).

It may be shown that the requirement that u be C^1 in S (in particular, that $\partial u/\partial n$ be continuous across the free boundary) is satisfied to within terms of order ε^2.

We remark in closing that free boundary problems for equations similar to (6.1) arise in plasma physics; in [1], for example, are considered equations of the form $Lu + f(u) = 0$, where L is an elliptic operator and f is, however, piecewise-linear in u, not discontinuous.

REFERENCES

1. G. Cenacchi, A. Taroni and A. Sestero. Numerical solution of the MHD equilibrium equation for an axially symmetric free-boundary plasma in a torus with arbitrary cross-section, Nuovo Cim. 25 B (1975) 279-294.

2. J. Chandra and B. A. Fleishman. Existence and comparison results for a class of nonlinear boundary value problems, Ann. Mat. Pura Appl. 101 (1974) 247-261.

3. B. A. Fleishman and T. J. Mahar. Boundary value problems for a nonlinear differential equation with discontinuous nonlinearities, Math. Balk. 3 (1973) 98-108.

4. B. A. Fleishman and T. J. Mahar. Boundary value problems with discontinuous nonlinearities: comparison of solutions, approximation, and continuous dependence on parameters, J. Diff. Eqs. (to appear).

5. B. A. Fleishman and T. J. Mahar. Analytic methods for approximate solution of elliptic free boundary problems, Nonlin. Anal. 1 (1977) 561-9.

6. B. A. Fleishman and Thomas J. Mahar. On the existence of classical solutions to an elliptic free boundary problem, Proceedings of the Third Scheveningen Conference on Differential Equations, eds. W. Eckhaus and E. M. de Jager (to appear).

7. L. E. Rubenštein. The Stefan Problem, Trans. Math. Monographs 27, Am. Math. Soc., Providence, R.I., 1971.

$$v(x) = \begin{cases} \cosh x + A \sinh x, & 0 \le x \le x_o \\ B \cosh(1-x), & x_o \le x \le 1 \end{cases}$$

where

$$A = B[(1/\bar{x}_o)\cosh x_o \cosh \bar{x}_o - \sinh 1],$$

$$B = [\cosh 1 - (1/\bar{x}_o)\sinh x_o \cosh \bar{x}_o]^{-1}, \qquad (6$$

$$\bar{x}_o = 1 - x_o.$$

Finally, assuming as in the previous problem that th perturbed free boundary has (up through terms of first ord ε) the form

$$x = x_o + \varepsilon g(y),$$

we substitute this for x in

$$u(x,y) = u_o(x) + \varepsilon v(x)\cos y = \mu$$

(neglecting terms of order ε^2) and find

$$g(y) = -(B/\bar{x}_o)\cosh \bar{x}_o \cos y. \qquad (6.1$$

Thus, for given $\mu \varepsilon (0, 1/4)$ and $0 < \varepsilon < \mu$, we have an approximate solution of $P_2(\varepsilon)$ of the form

$$u(x,y) = \begin{cases} x_o x + \varepsilon(\cosh x + A\sinh x)\cos y, & 0 \le x \le x_o + \varepsilon g(\\ x - \frac{1}{2}(x^2 + x_o^2) + \varepsilon B\cosh(1-x)\cos y, & x_o + \varepsilon g(y) \le x \le 1 \end{cases}$$

Bernard A. Fleishman and Ross Gingrich
Department of Mathematical Sciences
Rensselaer Polytechnic Institute
Troy, New York 12181

Thomas J. Mahar
Department of Mathematics
Utah State University
Logan, Utah 84321

Research supported by the U. S. Army Research Office under contract with Rensselaer Polytechnic Institute.

MOVING BOUNDARY PROBLEMS

HODOGRAPH METHODS AND THE SMOOTHNESS OF THE FREE BOUNDARY IN THE ONE PHASE STEFAN PROBLEM

David Kinderlehrer and Louis Nirenberg

1. INTRODUCTION

We discuss the smoothness of a free boundary arising in a parabolic variational inequality as for example in a one phase Stefan problem.

A one phase Stefan problem is the description, typically, of the melting of a body of ice maintained at $0^{\circ}C$ in contact with a region of water. The unknowns of the problem are the interface which separates the regions of water and ice and the temperature distribution of the water. The free boundary we study is identical to this interface when the variational inequality is the reformulation of the Stefan problem suggested by G. Duvaut [4]. We limit ourselves to a problem where some initial smoothness of the associated free boundary has already been achieved.

It is worthwhile noting that the interface separating the ice and water regions at a given time t depends on the data of the problem prescribed only up to time t; hence, it cannot be expected to depend analytically on the time variable. This notwithstanding, it exhibits rather precise behavior: it depends analytically on the space variables and is in the second Gevrey class with respect to the time variable. A function $\phi(t)$ in $(0,1)$ is in the second Gevrey class if for every compact subinterval there are positive constants C_0, M such that on the interval

$$|\partial_t^j \phi| \leq C_0 M^j (j!)^2 , \quad j = 0,1,2,\ldots$$

ISBN 0-12-757350-X

or, with different constants,

$$|\partial_t^j \phi| \leq C_0 \, M^j (2j)! \; , \quad j = 0,1,\ldots$$

Furthermore the free boundary is analytic in all variables during any time interval during which the heat supply everywhere is analytic. In the case of one space dimension, where the free surface is a curve, this result is due to A. Friedman [5].

Other work on the infinite differentiability of the free boundary in the one-dimensional Stefan problem is due to Cannon and Hill [3] and D.Schaeffer [12]. In deriving initial regularity we make use of a theorem of L. Caffarelli [1],[2] which allows us to pass at a certain point from a Lipschitz free boundary to a continuously differentiable one.

We also provide some suggestions for the two phase problem.

Our method consists of introducing appropriate hodograph and Legendre transformations which will have the effect of "straightening" the free boundary at the expense of replacing the equation by a highly non-linear one. For the solution of the non-linear equation we are able to demonstrate a precise regularity theorem. Proofs of the assertions given here may be found in [7], [8], [9].

The precise nature of our conclusions may be illustrated by this theorem. Consider a solution of the inhomogeneous heat equation

$$(1.1) \qquad -\Delta u + u_t = -a \quad \text{in } \Omega$$

where $\Omega \subset R^{n+1}$ is a certain domain whose boundary contains a hypersurface Γ with the property

$$(1.2) \qquad u = u_{x_i} = 0 \quad \text{on } \Gamma \, , \quad 1 \leqq i \leqq n \; .$$

Theorem A [8]. Suppose that u satisfies (1.1), (1.2) and that $u, u_{x_i} \in C^1(\Omega \cup \Gamma)$, $i = 1,\ldots,n$. Assume Γ admits the representation

$$\Gamma: x_n = g(x',t)\ ,\ |x'| < 1\ ,\ 0 < t < T\ ,$$

with $g \in C^1$. Assume furthermore that $a(x,t) > 0$ is an analytic function of x and of second Gevrey class in t; more precisely, that for some positive constants C, M and all derivatives

$$\|\partial^\alpha \partial_t^k a\|_{L^\infty(\Omega)} \leqq C M^{2k+|\alpha|}(2k+|\alpha|)\,! \tag{1.3}$$

Then u and g are analytic functions of the space variables x and in the second Gevrey class with respect to t in a neighborhood of any point of $\Omega \cup \Gamma$.

Above, $\alpha = (\alpha_1,\ldots,\alpha_n)$ is a multi-index and $\partial^\alpha = \partial_1^{\alpha_1} \ldots \partial_n^{\alpha_n}$, $\partial_i = \partial/\partial x_i$, $\partial_t = \partial/\partial t$. Also, $x' = (x_1,\ldots,x_{n-1})$.

We point out that the hypotheses of Theorem A may be verified in some important cases [6]. The passage from a Lipschitz free boundary to a continuously differentiable one is make possible by a result of Caffarelli [1], [2].

For $a \equiv 1$, $u(x,t)$ is the solution of the one phase Stefan problem as formulated by Duvaut [4] (cf. also [6].) The temperature of the water distribution is given by

$$\Theta(x,t) = u_t(x,t)\ ,\quad (x,t) \in \Omega\ .$$

No result better than Theorem A need hold even in this case; namely, u need not lie in any better Gevrey class ([8] §5.)

A second application of our method is a global analyticity theorem, which we formulate in terms of the classical Stefan problem. Introduce the polar coordinates in $\mathbb{R}^n$

$$\rho = |x|$$
$$\omega = (\omega_1, \dots, \omega_{n-1}) \in \mathbb{T} \text{ with}$$
$$\mathbb{T} : -\pi < \omega_i \leqq \pi , \ i < n-1 , \ 0 \leqq \omega_{n-1} < \pi \quad \text{if} \quad n > 2 ,$$
$$\mathbb{T} : -\pi < \omega \leqq \pi \quad \text{if} \quad n = 2$$

and assume that Γ is a surface representable in the form

$$\text{(1.4)} \qquad \Gamma: \rho = g(\omega,t) , \omega \in \mathbb{T}, \ 0 < t < T , \quad \text{with} \quad 1 < g(\omega,t) \in C^\infty(\mathbb{T} \times (0,T)) .$$

Suppose that $\Theta(x,t)$ and Γ satisfy

$$\text{(1.5)} \qquad \begin{cases} -\Delta\Theta + \Theta_t = 0 \\ \Theta > 0 \end{cases} \quad \underline{\text{in}} \ D = \{(x,t): 1 < |x| < g(\omega,t) , \ \omega \in \mathbb{T}, \ 0 < t < T\}$$

$$\text{(1.6)} \qquad \begin{cases} \Theta_x^2 \equiv \Theta_\rho^2 + \frac{1}{\rho^2} \Sigma \Theta_{\omega_i}^2 = \Theta_t \\ \Theta = 0 \end{cases} \quad \text{on } \Gamma$$

$$\text{(1.7)} \qquad \Theta = h \quad \text{on} \quad |x| = 1 , \ 0 < t < T ,$$

where $h(x,t)$ is given. The value of Θ when $t = 0$ is not relevant to our considerations.

<u>Theorem B [8]</u>. Let Θ,Γ satisfy (1.4), (1.5), (1.6), (1.7).

If $h(x,t)$ is positive and analytic, then

$$\Gamma : \rho = g(\omega,t) \ , \ \omega \in \ \ , \quad 0 < t < T \ ,$$

is an analytic hypersurface.

As we have remarked, $\Theta = u_t$ where u is a solution of (1.1), (1.2) with $a \equiv 1$, once Γ is assumed, say C^1, its C^∞ nature is assured by <u>Theorem A</u>.

2. PARTIAL HODOGRAPH AND LEGENDRE TRANSFORMATIONS

Let us assume for simplicity the hypotheses of Theorem A and that $0 \in \Gamma$ with the x_n direction normal to Γ at 0. We introduce the change of variables

$$(2.1) \qquad \begin{aligned} y_\alpha &= x_\alpha \qquad 1 \leqq \alpha \leqq n-1 \\ y_n &= -u_n \\ \tau &= t \end{aligned} \qquad ,$$

$(u_j = u_{x_j}$, etc) and the function

$$(2.2) \qquad v(y,\tau) = x_n y_n + u(x,t) \quad .$$

We refer to (2.1) as a partial hodograph transformation and to v in (2.2) as the Legendre transform of u. Note that (2.1), (2.2) differ from the customary definitions by a change in sign. It is easy to see that (2.1) is 1:1 near $(x,t) = (0,0)$ in Ω. First note that since $u = u_i = 0$ on Γ, $u_t = -u_n g_t = 0$ on Γ so

$$\Delta u = a > 0 \quad \text{on } \Gamma \ .$$

In addition, since $(0,\dots,0,1,0)$ is normal to Γ at $(0,0)$

$$u_{i\alpha}(0,0) = 0, \ 1 \leqq i \leqq n, \quad 1 \leqq \alpha \leqq n-1 ,$$

so

$$u_{nn}(0,0) = a(0,0) > 0 \quad .$$

This implies $\partial(y,\tau)/\partial(x,t)$ is non-singular at $(0,0)$. Under the mapping (2.1), a neighborhood of $(x,t) = (0,0)$ in Ω is mapped, say, onto a set

$$U \subset \{(y,\tau) : y_n > 0\}$$

and a neighborhood of $(x,t) = (0,0)$ in Γ is mapped onto

$$S \subset \{(y,\tau) : y_n = 0\} \ .$$

The property of the Legendre transform is that

$$\begin{aligned} dv &= x_n\, dy_n + y_n\, dx_n + du \\ &= x_n\, dy_n + \sum_{\alpha<n} u_\alpha\, dy_\alpha + u_t\, d\tau \end{aligned}$$

or

$$\text{(2.3)} \qquad \begin{aligned} v_n &= x_n \\ v_\alpha &= u_\alpha \, , \qquad 1 \leqq \alpha \leqq n-1 \ . \\ v_\tau &= u_t \end{aligned}$$

In particular a portion Γ' of Γ admits the representation

(2.4) $$\Gamma': x_n = \frac{\partial v}{\partial y_n}(x_1,\dots,x_{n-1},0,t), \quad (x_1,\dots,x_{n-1},0,t) \in S .$$

The smoothness of Γ' has become a question of that of v in $U \cup S$. One easily computes from (2.3), setting $y' = (y_1,\dots,y_{n-1})$, that v is a solution to the non-linear parabolic problem

(2.5)
$$-\frac{1}{v_{nn}} - \frac{1}{v_{nn}} \sum_{\alpha<n} v_{\alpha n}^2 + \sum_{\alpha<n} v_{\alpha\alpha} - v_\tau - a(y', v_n, \tau) = 0 \text{ in } U$$
$$v = 0 \text{ on } S.$$

The conclusion of Theorem A now follows from Theorem 1 of [9].

We employ a different transformation in the proof of Theorem B. We give a sketch of this supposing that $h(x,t) \equiv 1$ for simplicity. In this case we choose new independent variables

(2.6)
$$y_i = (1 + c_0 e^{-c_1\rho} u(x,t)) \frac{x_i}{|x|} \qquad (x,t) \in D$$
$$\tau = t$$

or in polar coordinates

(2.6')
$$r = 1 + c_0 e^{-c_1\rho} u(x,t)$$
$$\eta_i = \omega_i, \quad 1 \leq i \leq n-1, \qquad (x,t) \in D .$$
$$\tau = t$$

For an appropriate choice of c_0, c_1, it may be shown that all of D is mapped 1:1 onto the cylinder

$$G = \{(y,\tau): 1 < |y| < 2, \quad 0 < \tau < T\}$$

and the image of Γ is the cylindrical section

$$C = \{(y,\tau):\ |y| = 1\ ,\quad 0 < t < T\}\ .$$

We choose

$$(2.7)\qquad v(y,\tau) = v(r,\eta,\tau) = \rho\ ,\qquad (y,\tau) \in G\ ,$$

to be our new dependent variable.

One may verify that $v(y,\tau)$ is a solution of the problem

$$(2.8)\qquad -F(D^2v,Dv,v,y,\tau) + v_\tau = 0 \qquad\qquad \text{in } G$$

$$(2.9)\qquad\qquad v_\tau = \Phi(Dv,v,y,\tau),\ \rho=1,\ 0<t<T$$

$$(2.10)\qquad\qquad v = 1 \qquad\qquad \rho=2,\ 0<t<T$$

where $(F_{v_{ij}})$ is positive definite for the appropriate range of variables. Also, F and Φ are analytic functions of their arguments.

The condition (2.9) is of a mixed type which does not fit into the usual theory of coercive boundary conditions of Ladyzenskaya, Solonnikov, Ural'ceva [11]; however in the present instance

$$(2.11)\qquad \frac{\partial\Phi}{\partial v_\rho} \equiv \sum_1^n \frac{\partial\Phi}{\partial v_j}x_j > 0 \quad \text{for}\ \ \rho = 1\ ,\quad 0<t<T\ .$$

The property (2.11) leads to an L^2 estimate for (2.8), (2.9), (2.10) which in turn permits the conclusion that $v(y,\tau)$ is analytic in (y,τ) for $1 \leqq \rho \leqq 2\ ,\ \ 0<\tau<T$.

3. THE TWO PHASE STEFAN PROBLEM

The two phase Stefan problem is also amenable to the formulation we have given in §2. Let us consider a local formulation in a single space variable x. Suppose that $\Gamma: x = s(t)$, $0 < t < 1$, is a C^1 curve, $s(0) = s_0 \in (0,1)$, and that Θ satisfies

$$(3.1) \qquad \begin{aligned} -\Theta_{xx} + \Theta_t &= 0 \\ \Theta &= 0 \end{aligned} \qquad \text{for } 0 < x < s(t), \quad 0 < t < 1$$

$$(3.2) \qquad \begin{aligned} -\Theta_{xx} + \Theta_t &= 0 \\ \Theta &< 0 \end{aligned} \qquad \text{for } s(t) < x < 1, \quad 0 < t < 1$$

$$(3.3) \qquad \begin{aligned} \Theta &= 0 \\ \Theta_x^+ - \Theta_x^- &= s' \end{aligned} \qquad \text{for } (x,t) \in \Gamma, \quad 0 < t < 1$$

Formally, we select new independent variables

$$(3.4) \qquad \begin{aligned} \xi &= \Theta \\ \tau &= t \end{aligned}$$

and a new dependent variable

$$(3.5) \qquad v(\xi,\tau) = x \ .$$

With an appropriate choice of data on the segments $x = 0$, $0 < t < 1$; $x = 1$, $0 < t < 1$; and $0 < x < 1$, $t = 0$ the transformation (3.4) may be seen to be one-to-one in the square $Q = \{(x,t): 0 < x < 1, 0 < t < 1\}$. In any event, the maximum principle implies that (3.4) is 1:1 in a neighborhood of any

$(x_0,t_0) \in \Gamma$. Under (3.4), the interface Γ is mapped onto the segment $0 < \tau < 1$ of the τ-axis. Now

$$d\Theta = \Theta_x dx + \Theta_t dt$$
$$= \Theta_x dv + \Theta_t d\tau$$

or

$$dv = \frac{1}{\Theta_x} d\xi - \frac{\Theta_t}{\Theta_x} d\tau \quad .$$

It follows that in the image of Q under (3.4)

$$(3.6) \qquad -\frac{v_{\xi\xi}}{v_\xi^2} + v_\tau = 0 \quad \text{for} \quad \xi \neq 0 \quad .$$

We next determine the boundary conditions satisfied by v on $\xi = 0$. First note that v is continuous on $\xi = 0$, so

$$(3.7) \qquad v^+ = v^- \ , \quad 0 < \tau < 1 \ , \quad \xi = 0 \quad .$$

Recalling that $\Theta_x = \frac{1}{v_\xi}$, it follows from (3.3) that

$$(3.8) \qquad \frac{1}{v_\xi^-} - \frac{1}{v_\xi^+} = s' = v_\tau \ , \quad 0 < \tau < 1, \quad \xi = 0 \quad .$$

The plus and minus signs have been interchanged because the water region, where $\Theta > 0$, corresponds to the right ξ plane although it lies to the left of Γ .

We now introduce a non-linear parabolic system. Define

$$(3.9) \qquad w(\xi,\tau) = v(-\xi,\tau) \quad \text{for} \quad \xi > 0 \quad .$$

Now v and w are both defined in $\xi > 0$, $0 < \tau < 1$. For definiteness, let us assume that both functions are defined in the domain $\{(\xi,\tau): 0 < \xi < 1, 0 < \tau < 1\}$. This may be achieved by a change in scale. We now see that

$$(3.10)\quad \begin{cases} -\dfrac{v_{\xi\xi}}{v_\xi^2} + v_\tau = 0 \\ \\ -\dfrac{w_{\xi\xi}}{w_\xi^2} + w_\tau = 0 \end{cases} \qquad \text{in } 0 < \xi < 1,\ 0 < \tau < 1$$

$$(3.11)\quad \begin{cases} v - w = 0 \\ \\ \dfrac{1}{v_\xi} + \dfrac{1}{w_\xi} = -v_\tau = -w_\tau \end{cases} \qquad \text{for } 0 < \tau < 1,\ \xi = 0 .$$

We shall state an energy estimate pertaining to this system. Although we have not checked the details, we believe this estimate to imply that the curve Γ is of $2^{\underline{nd}}$ Gevrey class in t and that it is analytic under appropriate assumptions about the initial temperature and the contributed heat.

Let $a(\xi,\tau)$, $b(\xi,\tau)$ be smooth positive functions in $\overline{Q}, Q = \{(\xi,\tau): 0 < \xi < 1, 0 < \tau < 1\}$, and suppose that $\overline{v}, \overline{w}$ satisfy

$$(3.12)\quad \begin{matrix} -a\,\overline{v}_{\xi\xi} + \overline{v}_\tau = f \\ \\ -b\,\overline{w}_{\xi\xi} + \overline{w}_\tau = g \end{matrix} \qquad \text{in } Q$$

$$(3.13)\quad \begin{matrix} \overline{v} - \overline{w} = 0 \\ \overline{v}_\tau = \overline{w}_\tau = \overline{v}_\xi + \overline{w}_\xi + \varphi \end{matrix} \qquad \text{for } \xi = 0,\ 0 < \tau < 1$$

(3.14) $\bar{v} = \bar{w} = 0$ for $\xi = 1,\ 0 < \tau < 1$ and $\tau = 0,\ 0 < \xi < 1$

where f, g, and φ are given functions. Then

$$\iint_Q (\bar{v}_{\xi\xi}^2 + \bar{v}_\tau^2 + \bar{w}_{\xi\xi}^2 + \bar{w}_\tau^2)\, d\xi\, d\tau + \int_{\xi=0} (\bar{v}_\tau^2 + \bar{w}_\tau^2)\, d\tau \leqq$$

$$\leqq C \iint_Q (f^2 + g^2 + \varphi^2 + \varphi_\xi^2 + \bar{v}^2 + \bar{w}^2)\, d\xi\, d\tau ,$$

where C depends on a, b and their first derivatives with respect to ξ.

For other considerations regarding systems, we refer to [10].

REFERENCES

1. L.A. Caffarelli, The smoothness of the free boundary in a filtration problem, Arch. Rat. Mech. and Anal.

2. _______________, The regularity of elliptic and parabolic free boundaries, Bull. AMS 82 (1976), 616-618.

3. J. Cannon and D. Hill, On the infinite differentiability of the free boundary in a Stefan problem, J. Math. Anal. and Appl. 22 (1968), 385-397.

4. G. Duvaut, Résolution d'un problème de Stefan (Fusion d'un bloc de glâce à zero degré) CRAS Paris 276 (1973), 1461-1463.

5. A. Friedman, Analyticity of the free boundary for the Stefan problem, Arch Rat. Mech. and Anal. 61 (1976), 97-125.

6. A. Friedman and D. Kinderlehrer, A one-phase Stefan problem, Ind. Math. J. 24 (1975), 1005-1035.

7. D. Kinderlehrer, and L. Nirenberg, Regularity in free boundary problems, Ann. S.N.S. - Pisa 4 (1977), 373-391.

8. ____________________, The smoothness of the free boundary in the one phase Stefan problem, C.P.A.M. (to appear).

9. ____________________, Analyticity at the boundary of solutions of nonlinear second order parabolic equations,

10. D. Kinderlehrer, L. Nirenberg, and J. Spruck (to appear).

11. O.A. Ladyzenskaya, V.A. Solonnikov, N.N. Ural'ceva, Linear and quasilinear equations of parabolic type, Moscow (1967) AMS translation vol. 23, 1968.

12. D. Schaeffer, A new proof of the infinite differentiability of the free boundary in the Stefan problem, J. Diff. Eq. 20 (1976), 266-269.

David Kinderlehrer
School of Mathematics
University of Minnesota
Minneapolis, MN

Louis Nirenberg
Courant Institute
New York University
New York, NY

Methods Papers

MOVING BOUNDARY PROBLEMS

THE NUMERICAL SOLUTION OF MULTIDIMENSIONAL STEFAN PROBLEMS - A SURVEY

Gunter H. Meyer

It is the purpose of this paper to give a brief survey over commonly used and some recent methods for the numerical solution of multi-dimensional Stefan and Stefan type problems.

1. STATEMENT OF THE PROBLEM

We shall choose as our model the Stefan problem of conductive heat transfer with change of state. For definiteness, we shall think of a two phase system such as water and ice occupying a given domain $D \subset \mathbb{R}_2$ with prescribed temperature or flux on the boundary ∂D of D. The classical Stefan problem requires the determination of the temperature field over D. The freezing isotherm in D, denoted here by $\Gamma(t)$, is a free surface which is not known a priori but must be found as part of the solution of the problem. For ease of notation it is assumed that D consists of two regions D_1 and D_2 occupied by ice and water, respectively, and separated by $\Gamma(t)$.

In D_i the temperature u_i has to satisfy the usual heat equation

$$L_i u_i \equiv \nabla \cdot k_i \nabla u_i - c_i \rho_i \frac{\partial u_i}{\partial t} = f_i, \ i = 1,2 \tag{1.1}$$

where the thermal properties and the source terms may be functions of space and time. In applications, these quantities may also depend on the temperature; however, linearization and iteration are frequently employed so that at least in each iteration equations of the form (1.1) occur.

Variations in the formulation of the Stefan problem are due

ISBN 0-12-757350-X

primarily to the way in which the phase change on $\Gamma(t)$ is expressed. Let us assume that the temperature is normalized so that freezing takes place at

$$u_1 = u_2 = 0 \tag{1.2}$$

and that $\Gamma(t) = \{(x_1,x_2) : u_1(x_1,x_2,t) = 0\}$. If $\Gamma(t)$ evolves smoothly as a function of time and space, and if $\Gamma(t)$ admits a normal everywhere in D then the heat flux balance on the free surface can be approximated by

$$k_1\nabla u_1 - k_2\nabla u_2 = \lambda\rho \frac{dR}{dt} \tag{1.3}$$

where $R(t) = (x_1(t),x_2(t))$ is the position vector of a point on $\Gamma(t)$. All derivatives are understood as one sided limits on $\Gamma(t)$, while $\lambda\rho$ is the actual latent heat per unit volume. We shall ignore here density and volume changes on $\Gamma(t)$ (which, for certain systems, may appreciably influence the movement of $\Gamma(t)$ [22]).

The equations (1.1,2,3) describe the classical Stefan problem. By the term "Stefan type problem" we mean free surface problems for general parabolic operators whose elliptic part is not necessarily of divergence type (allowing anisotropy and convection), and whose free surface conditions are given as general functions of u_1,u_2 and their gradients (allowing, for example, space dependent source terms on $\Gamma(t)$).

2. EXPLICIT VS. IMPLICIT NUMERICAL METHODS

The Stefan problem is a typical (nonlinear) evolution problem and, like standard parabolic problems, allows an explicit determination of the approximate solution at successive time levels. Fully explicit finite difference methods have been used repeatedly with good success for a number of solidification and ablation problems (e.g. [1], [9]). However, the explicit solution of diffusion equations imposes severe stability restrictions on the space-time resolution of the problem. As is well known the mesh ratios must obey the well known inequality

$$\frac{k\Delta t}{\rho c \Delta x^2} \leq \frac{1}{2n}$$

where n is the space dimension. If we now consider the solidification of a slab of zinc then this inequality becomes approximately [7]

$$\Delta t = 3\Delta x^2 \quad \frac{cm^2}{sec} \quad .$$

Hence space steps of the order of 1 cm can require time steps of the order of 3 seconds. A solidification process takes on the order of hours so that an explicit method may well be competitive. On the other hand, freezing of soil leads to an estimate of the magnitude

$$\Delta t \leq 0.04 \quad \frac{cm^2}{hr} \quad .$$

(sec, e.g. [8, p.103]); hence if the space step again is of the order of 1 cm then Δt is of the order of .04 hrs. In a typical application freezing may be monitored over years [23] so that explicit methods no longer appear attractive. An alternative to the fully explicit method is a semi-implicit technique (sometimes called Huber's method) where the location of $\Gamma(t)$ is predicted explicitly over some interval, and where the diffusion equation is then solved implicitly on the predicted domain. In effect this means integrating the differential equations on a known but time-dependent domain. This method applied to the classical Stefan problem is not subject to stability restrictions. As discussed in [7], the boundary predictions are self-correcting in that underpredicting the speed of the surface at a given time level leads to damped overpredicting at the next time level.

Semi-implicit methods will break down when $\Gamma(t)$ moves rapidly because of low latent heat, or when source terms on the free boundary accelerate an already overpredicted free boundary. Fully implicit methods, on the other hand, frequently still can cope. For example, Fig. 1a below shows the movement of the solid-liquid interface between a slab of zinc held at the phase change temperature and molten zinc initially 100°C above this temperature. Curve 1 is computed with the fully implicit sweep algorithm described in [10], curve 2 is based on the sweep method applied to a predicted domain. Since the liquid phase initially has a uniform temperature the first boundary prediction is $s'(0) = 0$. It is seen that a small error is initially introduced which quickly disappears. Both the fully and the semi-implicit methods performed equally well.

On the other hand, if the latent heat per unit volume is decreased from $\lambda = 26.6$ to $\lambda = 0.01$ (one may think of a zinc-like substance with traces of meltable zinc) then the boundary moves about 100 times faster and, as Fig. 1b shows, the semi-implicit method cannot recover from the poor initial prediction. A similar misbehavior would be expected from a fully explicit method. A fully implicit method, on the other hand, yields a free surface which remains virtually unchanged under further mesh refinements.

In summary, explicit methods are simple to apply but limited by stability constraints on the achievable space-time resolution, while implicit methods require the solution of diffusion equations. In standard applications such as the melting of ice or the solidification of metal the interfaces move relatively slowly; here the stability of the free surface location can exert itself, making extreme care in placing the free boundary unnecessary. Indeed, perturbing the boundary to pass through fixed mesh points, or solving the equations on a regular grid is frequently practiced. For these problems most numerical methods will work well as long as the diffusion equations are solved correctly and the boundary conditions are approximated consistently.

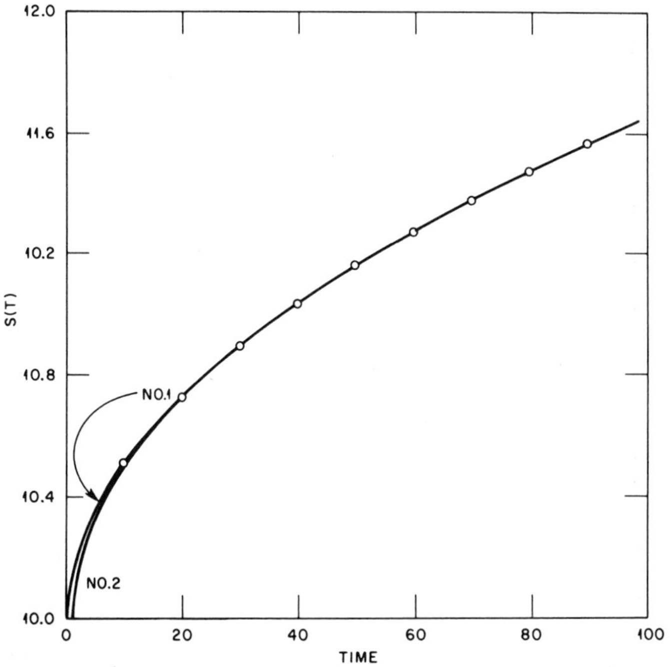

Fig. 1a. Free boundary as a function of time for melting zinc.
Curve 1: Fully implicit method
Curve 2: Semi-implicit method
$\Delta x = 12/500$ cm, $\Delta t = 1$ sec, 100 steps
15 sec Cyber 74 CPU time.

For non-standard problems with unusual boundary conditions and rapidly moving fronts implicit methods or an iterative use of explicit methods would appear to be required.

3. SOME SPECIFIC METHODS FOR STEFAN PROBLEMS

Several detailed surveys on the numerical solution of Stefan problems are available [19], [6], [12] so that the discussion here is only meant to complement and bring up to date the earlier accounts. One dimensional methods will be ignored although they may find use for multidimensional problems in connection with locally one dimensional methods.

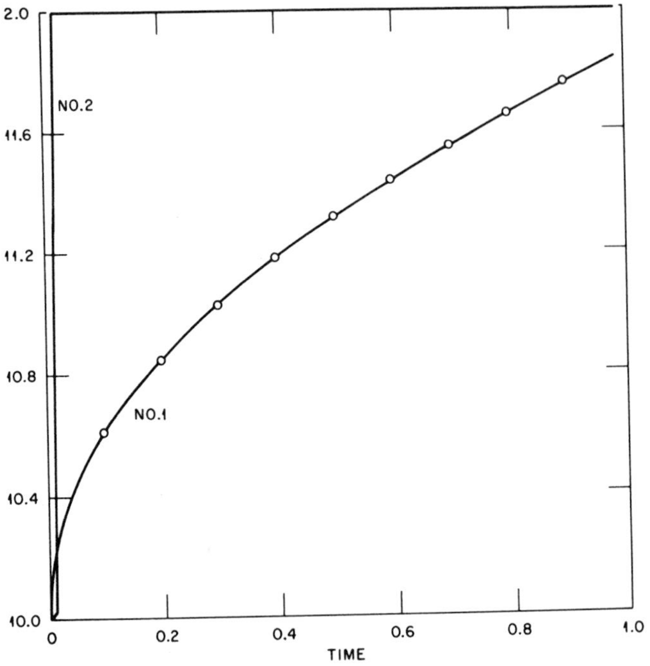

Fig. 1b. Free boundary as a function of time for a zinc-like substance with negligible latent heat
Curve 1: Fully implicit method
Curve 2: Semi-implicit method
$\Delta x = 12/500$ cm, $\Delta t = .01$ sec, 100 steps
15 sec Cyber 74 CPU time.

Various methods have been proposed for the numerical solution of phase transition problems. They differ primarily in the way that heat transfer on the phase boundary is modeled. If the flux balance (1.3) or related expressions are used then the position of $\Gamma(t)$ enters into the approximating equations and so called front tracking methods result. Such methods require that the phase front evolve smoothly in time and space, and one will usually have to draw on some a priori knowledge of the solution based on the physical model in order to judge whether the front is trackable. In a given application such decision is usually easy to make. Let us now look at some specific front tracking methods.

I. Front tracking methods. Much of the numerical work on Stefan problems has been based on finite difference approximations defined on a fixed rectangular grid. It is apparent from the flux condition (1.3) and the derived expression

$$k_1 \frac{\partial u_1}{\partial n} - k_2 \frac{\partial u_2}{\partial n} = \lambda \rho \frac{ds}{dt} , \tag{3.1}$$

where s(t) is the speed normal to $\Gamma(t)$, that mesh points will advance off grid lines as the free surface moves. Work on a fixed grid will require expensive and cumbersome two or three dimensional interpolation. As a result the expressions (1.3) and (3.1) have not been popular for work in two or more space dimensions. On the other hand, by tracking $\Gamma(t)$ with (3.1) no assumption about the direction of movement is made which restricts the applicability of alternative front tracking methods as described below. In view of the stability of semi-implicit methods discussed above, an adaptation of self triangulating finite element routines for elliptic boundary value problems (see, e.g. [20]) to the time discretized Stefan problem with explicit free surface prediction should be simple and effective for many technical problems. Such methods are becoming available now (see [14] for a modified finite element-continuous time approach and [2] for a space-time Galerkin method) however, their use has been restricted so far to simple model problems. As with most finite element methods computer implementation tends to be time consuming and costly.

For work on a rectangular grid, tracking the front along grid lines will simplify interpolation between grid points by making it one dimensional. A different flux condition on $\Gamma(t)$ now becomes useful. Suppose the surface $\Gamma(t)$ is given by $\phi(x,y,t) = 0$. If we differentiate along the path traced out by

$(x(t), y(t)) \in \Gamma(t)$ we obtain

$$\frac{d\phi}{dt} \equiv \langle \nabla\phi, \left(\frac{dx}{dt}, \frac{dy}{dt}\right)\rangle + \frac{\partial\phi}{\partial t} = 0 .$$

Substitution of (1.3) into this expression yields the equation

$$\langle \nabla\phi, k_1\nabla u_1 - k_2\nabla u_2\rangle + \lambda\rho \frac{\partial\phi}{\partial t} = 0 \qquad (3.2)$$

from which various expressions can be derived under certain assumptions on the shape of $\Gamma(t)$. Let us suppose, for example, that the front moves primarily in the y-direction and that we can write

$$\phi(x,y,t) \equiv y - \psi(x,t) = 0 .$$

Equation (3.2) now becomes

$$\langle \left(-\frac{\partial\psi}{\partial x}, 1\right), k_1\nabla u_1 - k_2\nabla u_2 \rangle - \lambda\rho \frac{\partial\psi}{\partial t} = 0 . \qquad (3.3)$$

It is advantageous to eliminate $\frac{\partial u_i}{\partial x}$ from this equation in order to relate the flux in the y-direction with the movement in the y-direction. We note that the directional derivative of u_i on $\Gamma(t)$ must vanish so that

$$\langle \nabla u_i, \left(1, \frac{\partial\psi}{\partial x}\right)\rangle = 0, \ i = 1,2 .$$

Solving for $\frac{\partial u_i}{\partial x}$ and substituting into (3.3) we finally obtain

$$\left(1 + \left(\frac{\partial\psi}{\partial x}\right)^2\right) \left(k_1 \frac{\partial u_1}{\partial y} - k_2 \frac{\partial u_2}{\partial y}\right) = \lambda\rho \frac{\partial y}{\partial t} \qquad (3.4)$$

A similar relation for the flux balance along x can be obtained by assuming ϕ to be representable as $\phi(x,y,t) \equiv x - y(y,t) = 0$. A somewhat different derivation of (3.4) may be found in [16].

The fully explicit method of [9] is based on (3.4) and analogous expressions along the remaining axes. That paper demonstrates clearly the effectiveness of the explicit method in modeling the melting of a prism of ice, but also points out the complexity of the approximating equations near the interface and the need for a special starting solution at $t = 0$ if accuracy on a coarse mesh (necessary for stability) is to be maintained.

Equation (3.4) in partially inverted form is used in the isotherm migration method [3] where $u = u(x,y,t)$ is expressed as $y = y(u,x,t)$ and where a (nonlinear) differential equation for y is derived which is solved time explicitly on a fixed (u,x)-grid. Again, a special starting procedure is required which becomes critical because u must be and remain an invertible function of y.

Also based on (3.4) are the locally one dimensional methods presented in [11], [12], and [13] for Stefan type problems. When fully discretized they provide an implicit analog to the method of [9]. To illustrate the performance of the two methods examined so far let us apply the method of fractional steps as given in [11] and the method of lines as given in [12] to the melting of an ice prism as described as described in [3], [9], and [17]. The prism is thought to occupy the square $D = [0,2] \times [0,2]$. It is initially ice at the phase change temperature and then suddenly uniformly heated on all sides. Heat flow in the water phase D_2 between the wall and the ice is approximated by

$$u_{xx} + u_{yy} - u_t = 0 \qquad (x,y,)\varepsilon D_2$$

$$u = 1 \qquad (x,y)\varepsilon\partial D$$

$$u = 0 \qquad (x,y)\varepsilon\Gamma(t)$$

$$\frac{\partial u}{\partial n} = -\beta \frac{ds}{dt}, \text{ where } \beta = 1.5613 .$$

Because of symmetry we can work on the unit square.

If the flux condition is written as $(1 + (\frac{\partial x}{\partial y})^2)\frac{\partial u}{\partial x} = -\beta \frac{\partial x}{\partial t}$ or $(1 + (\frac{\partial y}{\partial x})^2) \frac{\partial u}{\partial y} = -\beta \frac{\partial y}{\partial t}$ and if the method of fractional steps is applied exactly as in [11] then starting with the Karman-Pohlhausen solution given in [17] for t = 0.0461 the melting fronts of Fig. 2 are obtained. Although these curves appear correct they differ from the isotherm migration results presented for the same initial condition in [3]. Along the axes the difference increases linearly with time from 0 to 15%. An attempt was made to apply instead a standard ADI method based on the fractional splitting [24, p.16]

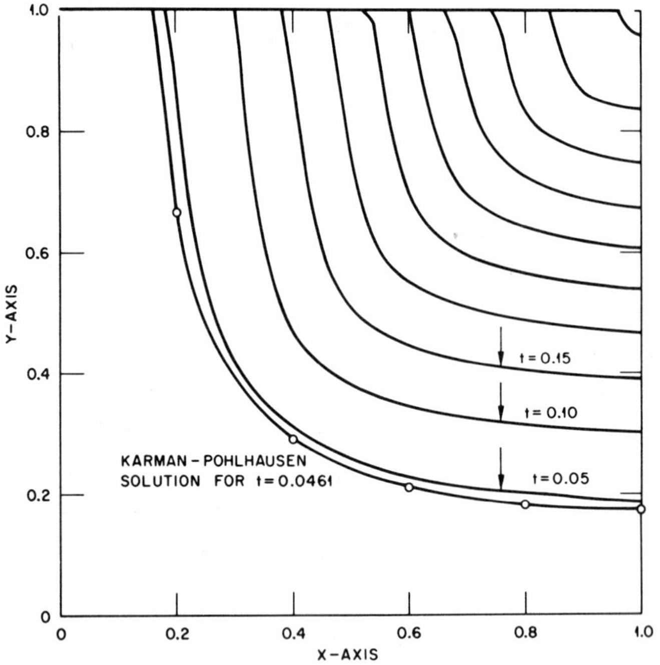

Fig. 2. Fractional step solution for melting ice. $\Delta x = \Delta y = 1/50$; free surface plotted every 10 time steps 3 min Cyber 74 CPU time.

$$u_{xx}^{n+1/2} - u_t^{n+1/2} = -u_{yy}^{n}$$

$$u_{yy}^{n+1} - u_t^{n+1} = -u_{xx}^{n+1}$$

where the source terms were obtained from the preceding time level by central differencing. As Fig. 3 shows the result was unstable.

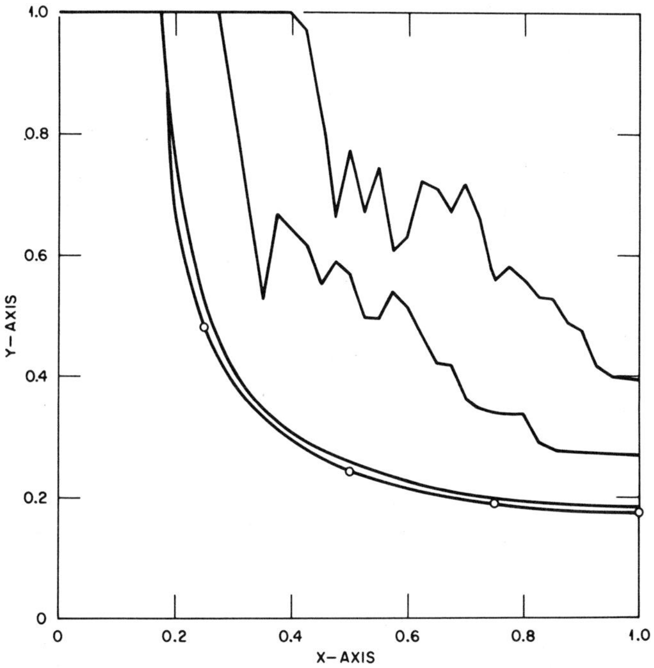

Fig. 3. Alternating direction fractional step method was unstable.

Thus the finely tuned balancing of the horizontal and vertical sweeps of the ADI method on irregular domains is lost when a free surface is present. Altogether, the performance of fractional step methods for this type of problem is not yet satisfactory, particularly when the end points of the free boundary are not pinned down by Dirichlet data.

The same problem then was solved with the method of lines as given in [12]. This time symmetry about the line x = y was used to minimize execution time. Starting again with the Karman-Pohlhausen solution used in [17], virtually identical results to those of [3] were obtained. However, since the method is implicit it may be started at t = 0 without additional work. As Fig. 4 shows the Karman-Pohlhausen solution, which is based on an assumed shape of the surface, turns out to be none too exact. This discrepancy already is noted in [9]. However, as the computation progresses the difference to the isotherm migration solution diminishes (Fig. 5) which is again an indication of the stability of the free surface. Thus, we see that the method of lines for the time discretized Stefan problem is a viable fully implicit numerical method.

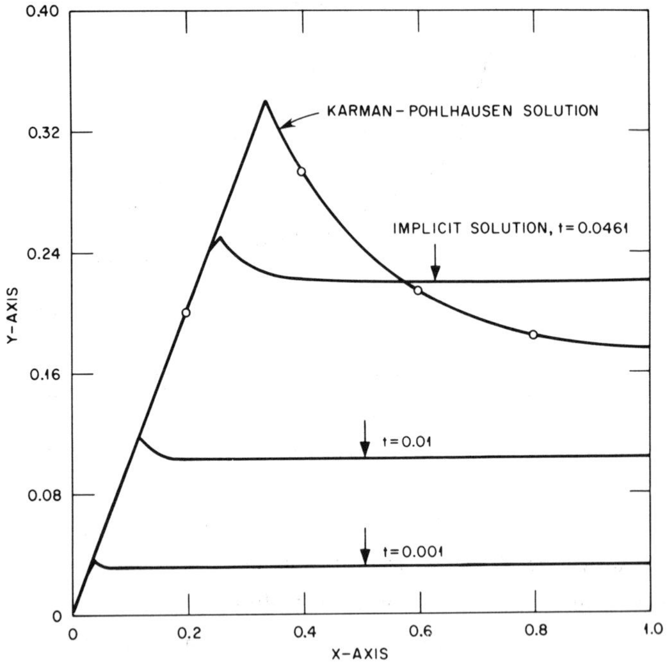

Fig. 4. Method of lines solution for the melting ice prism. $\Delta x = .02$, $\Delta y = 0.005$; free surface shown after each 10 time steps. 3 min. Cyber 74 CPU time

Finally, we note that the representation $\phi(x,y,t) \equiv t - \psi(x,y)$ leads to the interface condition

$$\langle \nabla\psi, k_1 \nabla u_1 - k_2 \nabla u_2 \rangle = \lambda\rho$$

which is frequently the starting point for analytic work on Stefan problems (see, e.g. [4]).

It is apparent from the above discussion that few modifications of the equations are required to handle more general interface conditions of the form

$$g_j(u_1, u_2, \nabla u_1, \nabla u_2, \frac{dR}{dt}, R) = 0, \quad j = 1,2,3$$

which will include a number of free surface dependent interface phenomena (for a discussion of one dimensional Stefan type problems see [10]). This freedom of changing the model without voiding the front tracking technique is considered a major advantage of this approach. Moreover, when coupled with time implicit locally one dimensional methods general source terms like (3.5) often are as simply accounted for as standard Stefan flux conditions [10], [13]. This freedom is lost when the interface conditions no longer appear explicitly, as in the methods discussed below.

II. Fixed domain methods. Even in classical heat transfer problems with change of phase the free surface simply may not be trackable. It may have kinks and corners, disappear and reappear, and in general lack all the smoothness required for a front tracking algorithm. Stefan problems with non-smooth as well as smooth free surfaces can be solved quite effectively by so-called fixed domain methods where the flux condition on the free surface is absorbed into the differential equations which then are solved on the fixed domain D. The surface $\Gamma(t)$ is obtained a posteriori by tracing the contour $u = 0$.

The most commonly used and studied method of this kind is

based on the enthalpy transformation

$$H(u) = \int_{u_0}^{u} c(v)dv \tag{3.6}$$

where c(u) is the specific heat of the medium at temperature u, and where u_0 is an arbitrary reference temperature. For a typical two phase system with constant thermal properties the function H is piecewise linear and exhibits a jump proportional to the latent heat at the phase change temperature.

In terms of (3.6) the equations (1.1-3) can be rewritten as

$$\nabla \cdot k(u)\nabla u - \frac{\partial}{\partial t}\left(H(u)\right) = 0, \; x \varepsilon D \tag{3.7}$$

subject to the given boundary conditions on ∂D. Because of the irregularity of k and H at u = 0 this equation holds in the classical sense in D_1 and D_2 and in the distributional sense on D.

A number of time implicit and explicit techniques based on iterative methods for nonlinear parabolic equations have been described and analyzed to solve (3.7) and we refer to [12] and [21] for details and further references. It may also be noted that the enthalpy formulation can be modified to incorporate volume changes [22].

A fairly recent enthalpy related method is the so-called alternating phase truncation method [18] where the solution is advanced in fractional time steps by computing heat flow in the liquid and solid phases alternately. For a special one dimensional problem this method (assuming no spacial discretization errors) is shown and observed to be $O(\Delta t \ \ell n \frac{1}{\Delta t})^{1/2}$ which compares poorly with the observed $O(\Delta t^2)$ of the finite element method of [2] and the proven $O(\Delta t^{1/2})$ result of [5] for Huber's semi-implicit method.

Most recent work on fixed domain methods for free surface problems has concentrated on variational inequalities. To remove the singularity from (3.7) the so-called freezing index

$$\theta(x,t) = \int_0^t u(x,T)\,dT$$

is defined. Equation (3.6) is then integrated with respect to time leading to

$$\nabla \cdot k_1 \nabla\theta - c\rho \frac{\partial\theta}{\partial t} + \rho H(u(x,0)) = \begin{cases} 0 \text{ if } x \varepsilon D_1 \text{ at } t \\ \rho\lambda \text{ if } x \varepsilon D_2 \text{ at } t. \end{cases} \tag{3.8}$$

The boundary conditions are integrated as well and the resulting boundary value problem for (3.8) is cast into the form of a variational inequality defined on D which can be solved numerically as a constrained optimization problem. This approach promises to yield accurate numerical methods with precise error bounds [5], [15].

On the basis of experience gained so far, numerical methods based on a time implicit solution of the enthalpy equation (3.7) can be recommended highly for the solution of two and multi phase Stefan problems in several dimensions. The model equation is close to the physics of the heat transfer process and its solution can be based on standard finite difference or element methods for non-linear elliptic differential equations. Experience with variational inequalities is still limited.

The disadvantage of all fixed domain methods is their restricted applicability to problems where the boundary data on ∂D completely drive the heat transfer process in D. If external sources appear as in (3.5) which depend on the location of $\Gamma(t)$ then an a posteriori determination of the freezing isotherm no longer appears possible. In fact, it is not always clear how free surface conditions like (3.5) can be incorporated into an enthalpy formulation or what transformation of the dependent variable will lead to a variational inequality.

REFERENCES

1. D. R. Atthey, A Finite Difference Scheme for Melting Problems, J. Inst. Math. Appl., 13, 1974, pp. 353-366.

2. R. Bonnerot and P. Jamet, A Second Order Finite Element Method for the One-Dimensional Stefan Problem, Int. J. Num. Meth. Engng., 8, 1974, pp. 811-820.

3. J. Crank and R. S. Gupta, Isotherm Migration Method in Two Dimensions, Int. J. Heat Mass Transfer, 18, 1975, pp. 1101-1107.

4. G. Durant, The Solution of a Two Phase Stefan Problem by a Variational Inequality in Moving Boundary Problems in Heat Flow and Diffusion, J. R. Ockendon and W. R. Hodgkins, edts., Clarendon Press, 1975.

5. A. Fasano and M. Primicerio, Convergence of Huber's Method for Heat Conduction Problems with Change of Phase, ZAMM, 53, 1973, pp. 341-348.

6. L. Fox, What are the Best Numerical Methods, in Moving Boundary Problems in Heat Flow and Diffusion, J. R. Ockendon and W. R. Hodgkins, edts., Clarendon Press, 1975.

7. A. Huber, Über das Fortschreiten der Schmelzgrenze in einem linearen Leiter, ZAMM, 19, 1939, pp. 1-20.

8. A. R. Jumikis, Thermal Soil Mechanics, Rutgers University Press, 1966.

9. A. Lazaridis, A Numerical Solution of the Multidimensional Solidification (or Melting) Problem, Int. J. Heat Mass Transfer, 13, 1970, pp. 1459-1477.

10. G. H. Meyer, One Dimensional Parabolic Free Boundary Problems, SIAM Review, 19, 1977, pp. 17-33.

11. G. H. Meyer, An Alternating Direction Method for Multidimensional Parabolic Free Surface Problems, Int. J. Num. Meth. Engng., 11, 1977, p. 741-752.

12. G. H. Meyer, The Numerical Solution of Stefan Problems with Front Tracking and Smoothing Methods, J. Appl. Math. Comp., to appear.

13. G. H. Meyer, An Application of the Method of Lines to Multidimensional Free Boundary Problems, J. Inst. Math. Appl., to appear.

14. M. Mori, Stability and Convergence of a Finite Element Method for Solving the Stefan Problem, Publ. RIMS, Kyoto Univ., 12, 1976, 539-563.

15. J. T. Oden, Finite Element Methods for Certain Classes of Free Boundary Value Problems in Mechanics, these proceedings.

16. P. D. Patel, Interface Conditions in Heat Conduction Problems with Change of Phase, AIAA J, 6, 1968, p. 2454.

17. G. Poots, An Approximate Treatment of a Heat Conduction Problem Involving a Two-Dimensional Solidification Front, Int. J. Heat Mass Transfer, 5, 1962, pp. 339-348.

18. J. C. W. Rogers, A. E. Berger and M. Ciment, The Alternating Phase Truncation Method for the Numerical Solution of a Stefan Problem, to appear.

19. L. I. Rubinstein, The Stefan Problem, A. Solomon, transl., Amer. Math. Soc., 1971.

20. E. G. Sewell, An Adaptive Program for the Solution of Div(P(x,y)Gradu) = F(x,y,u) on a Polygonal Region, in Proc. of 2nd Finite Element Conference, J. R. Whiteman, edt., Academic Press, 1976.

21. N. Shamsundar, Comparison of Numerical Methods for Diffusion Problems with Moving Boundaries, these Proceedings.

22. N. Shamsundar and E. M. Sparrow, Effect of Density Change on Multidimensional Conduction Phase Change, J. Heat Transfer, 1976, pp. 551-557.

23. J. A. Wheeler, Permafrost Thermal Design for the Trans-Alaska Pipeline, these Proceedings.

24. N. N. Yanenko, The Method of Fractional Steps, M. Holt, transl., Springer Verlag, 1971.

Gunter H. Meyer
School of Mathematics
Georgia Institute of Technology
Atlanta, Georgia 30332

Research supported in part by the U.S. Army Research Office under Grant DA-AG29-76-G-0261

MOVING BOUNDARY PROBLEMS

THE INTERRELATION BETWEEN MOVING BOUNDARY PROBLEMS AND FREE BOUNDARY PROBLEMS

Colin W. Cryer

We discuss some of the interrelationships between moving boundary problems (parabolic and time dependent) and free boundary problems (elliptic and steady state): (i) "New" moving boundary problems suggested by free boundary problems; (ii) The use of numerical methods for moving (free) boundary problems to solve free (moving) boundary problems; (iii) Free boundary problems with unstable solutions or non-unique solutions.

1. INTRODUCTION

We define a <u>free boundary problem</u> (FBP; plural, FBPS) to be a boundary value problem involving differential equations on domains, parts of whose boundaries, the <u>free boundaries</u> (FB; plural FBS), are unknown and must be determined as part of the solution. A <u>moving boundary problem</u> (MBP; plural, MBPS) is defined to be an initial - value problem for differential equations involving unknown free boundaries.

Most FBPS can be thought of as the limit (for time t tending to infinity) of corresponding MBPS. For example, if we open a water faucet (or tap) then we obtain a MBP with the air/water interface as a FB, but after a short time the motion becomes steady and can be formulated as a FBP.

In this lecture we will discuss some of the ramifications of this interrelationship between MBPS and FBPS. Unless otherwise stated, when we speak of the limit of a MBP we will be referring to the limit for time t tending to infinity. An n-dimensional problem will be a problem involving n space dimensions. Thus, a 1-dimensional Stefan problem will involve one space dimension and time; we note that in the literature on MBPS, such a problem is often said to be two-dimensional.

ISBN 0-12-757350-X

2. APPLICATIONS OF FBPS AND MBPS

One of the purposes of this meeting is to provide opportunities for the cross-fertilization of ideas and the suggestion of new problems. In this connection, we have recently completed a bibliography of FBPS (Cryer [6]), and Table 1 shows the approximate distribution of the references in this bibliography among the various areas of application.

Area of application	Percentage of references
Fluid mechanics	58
Porous flow	10
Elasticity and plasticity	15
Heat conduction and diffusion	1
Electromagnetism	5
Gravitation	2
Chemical reactions	1
Coupled fields	5
Control theory	1
Optimization	1
Mathematical generalizations	1
	100

Table 1. Distribution of references on FBPS according to area of application.

Participants at this meeting may be surprised that only approximately 1% of the references in the bibliography are concerned with heat conduction and diffusion. There are several reasons for this:

(i) Much of the literature on MBPS in heat conduction and diffusion is concerned with one-dimensional problems, because it is quite difficult to handle even one space dimension. However, if the limit of a one-dimensional MBP does exist then it corresponds to a one-dimensional FBP which is a relatively trivial problem.

(ii) Many interesting and difficult MBPS have uninteresting limit solutions. For example, if we fill a thermos flask with hot water and drop in an ice cube, we observe a very complex MBP, but in the limit we have a flask full of warm water.

The reason for giving Table 1 is not to show how few FBPS arise in the areas of heat conduction and diffusion, but, rather, to emphasize the immense variety of FBPS, almost all of which give rise to MBPS. We will briefly discuss two such MBPS, chosen more-or-less at random.

<u>Problem 1</u>. <u>Hydrofoil craft</u>. At high speed, a hydrofoil craft (Figure 1) is supported above the water level by the forces acting on its hydrofoils, in the same way that an aircraft is supported by the forces acting on its airfoils (wings). Acosta [1] gives a useful survey of the subject. The main advantage of hydrofoil craft is that they can operate at speeds of up to 50 knots, which is much faster than the speeds attainable by normal surface craft. For this reason, hydrofoil craft have considerable advantages for passenger services and naval operations.

As shown in Figure 1, the motion of a hydrofoil craft involves two types of FBS: (i) the water/air interface between the water and the atmosphere; (ii) the water/water-vapor interfaces which separate the water from the water-vapor cavities which form at high speeds behind the hydrofoils. The vapor cavity behind a hydrofoil is often analysed as if it were stationary, but in fact cavitation is a high-speed phenomenon, the study of which requires the use of cameras capable of taking between 10^4 and 10^7 pictures per second. (Knapp, Daily, and Hammitt [12, p. 41]). The MBP describing the motion of a hydrofoil craft is thus an extremely difficult problem.

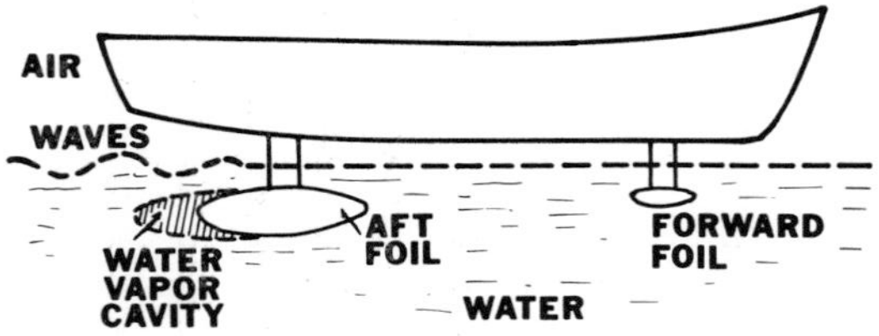

Figure 1. A hydrofoil craft. For clarity, the hydrofoils are not drawn to scale

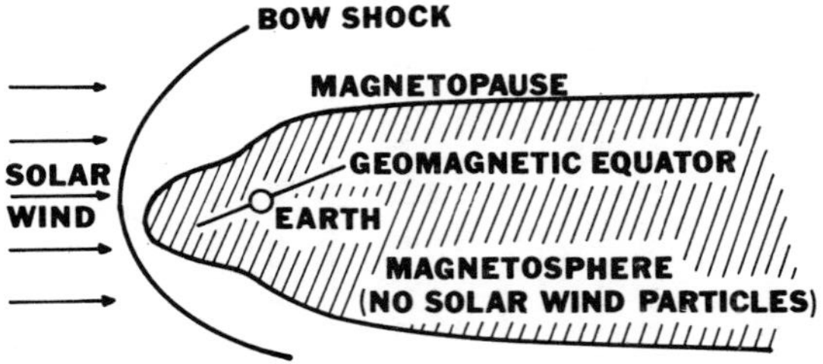

Figure 2. The earth's magnetosphere

Problem 2. The earth's magnetosphere. A stream of charged particles, called the solar wind, leaves the surface of the sun and bombards the earth (Figure 2). The particles are deflected by the earth's magnetic field, and are excluded from a region around the earth called the magnetosphere. The boundary of the magnetosphere, which is of course a FB, is known as the magnetopause and has been observed experimentally with the aid of satellites. In the case of high-speed gas flow past a blunt body, a FB, called the bow shock, separates the supersonic and subsonic regions, and an analogous bow shock occurs ahead of the magnetosphere (see Figure 2).

There is a considerable literature on the FBP describing the magnetosphere (Cryer [6, appendix 3, p. 64]). This suggests the study of similar MBPS, such as the MBP describing the effect upon the magnetosphere of sudden changes in the solar wind due to sun spots or sun flares.

3. NUMERICAL METHODS

There is considerable interplay between the numerical methods used for FBPS and MBPS. On the one hand, if a MBP is solved using an implicit method, then one is in effect solving a FBP at every time step. On the other hand, many FBPS have been solved numerically by formulating corresponding MBPS and then computing the solution until a steady-state is reached (Cryer [6, appendix 5, p. 63]). Here we wish to make a remark about this latter process.

If one solves a boundary-value problem numerically by solving an initial-value problem and computing until the solution becomes steady, then only the final solution is of interest and the intermediate results are of no value. It is, therefore, quite possible that the numerical calculations can be speeded up at the expense of the accuracy of the intermediate results. To our knowledge, this rather elementary remark has only been stated in the literature in a paper by Crocco [4] on the numerical solution of the steady Navier-Stokes equations.

The following example shows that very great improvements in speed can be attained. Consider the classical Dirichlet problem for Laplace's equation in a unit square R with boundary ∂R:

$$\nabla^2 u = 0, \quad \text{in } R; \; u = f \quad \text{on } \partial R . \tag{3.1}$$

If the problem is approximated using finite differences or piecewise linear finite elements one obtains a system of linear equations of which a typical equation is of the form

$$u_{i+1,j} + u_{i-1,j} + u_{i,j+1} + u_{i,j-1} - 4u_{i,j} = 0 . \tag{3.2}$$

If we choose to solve (3.1) using a time-dependent approach, then the most obvious approach is to replace the boundary-value problem (3.1) by the initial-value problem for the heat equation,

$$\begin{aligned} \frac{\partial v}{\partial t} = \nabla^2 v, \quad \text{in } R; \; v(\underline{x},t) &= f, \quad \text{for } \underline{x} \in \partial R; \\ v(\underline{x},0) &= g, \quad \text{for } \underline{x} \in R , \end{aligned} \tag{3.3}$$

where g is an estimate for the steady-state solution u. If (3.3) is approximated by a simple explicit finite difference formula, we obtain equations of the form

$$v_{i,j}^{(k+1)} = v_{i,j}^{(k)} + \frac{\Delta t}{(\Delta x)^2} \left[v_{i+1,j}^{(k)} + v_{i-1,j}^{(k)} + v_{i,j+1}^{(k)} + v_{i,j-1}^{(k)} - 4v_{i,j}^{(k)} \right] . \tag{3.4}$$

This numerical scheme is stable provided that $\Delta t/(\Delta x)^2 \leq 1/2$ (Forsythe and Wasow [7, p. 92]). Since we are interested in the limit we would usually take Δt as large as possible, namely $\Delta t = (\Delta x)^2/2$. With this choice of Δt the equations (3.4) take the form

$$v_{i,j}^{(k+1)} = \left[v_{i+1,j}^{(k)}+v_{i-1,j}^{(k)}+v_{i,j+1}^{(k)}+v_{i,j-1}^{(k)}\right]/4 \ , \tag{3.5}$$

which is identical with the method of simultaneous displacements (or Jacobi's method) for solving (3.2) (Forsythe and Wasow [7, p. 223]). If Δx is small then the spectral radius ρ_J corresponding to (3.5) is $\rho_J \doteq 1 - \pi^2(\Delta x)^2/2$ (Varga [14, p. 203]). We recall that for any iterative method with spectral radius ρ for solving (3.2), the number of iterations, m, required to reduce the error by a factor e is approximately $m = -1/\ell n\ \rho$ so that

$$m_J \doteq 2/\pi^2(\Delta x)^2 \ . \tag{3.6}$$

It is well known that the Jacobi method (3.5) is a poor method for solving Laplace's equation. It is much better to use successive over-relaxation,

$$v_{i,j}^{(k+1)} = v_{i,j}^{(k)} + \omega\left[v_{i+1,j}^{(k)}+v_{i-1,j}^{(k+1)}+v_{i,j+1}^{(k)}+v_{i,j-1}^{(k+1)}-4v_{i,j}^{(k)}\right] \ , \tag{3.7}$$

where the over-relaxation parameter ω is chosen so as to minimize the spectral radius ρ. The dependence of ρ upon ω is shown in Figure 3. For small Δx the optimal choice of ω is $\omega_{opt} \doteq 2 - 2\pi\ \Delta x$, (Varga [14, p. 203]) so that $\rho_{opt} = 1 - 2\pi\ \Delta x$, and

$$m_{opt} = 1/2\pi\ \Delta x \ . \tag{3.8}$$

Comparing (3.6) and (3.8) we see that the improvement in the rate of convergence is $m_J/m_{opt} \doteq 4/\pi(\Delta x)$ which is very large when Δx is small.

Garabedian [8] has observed that just as (3.5) can be interpreted as a numerical approximation to the heat equation (3.3), so (3.7) with $\omega = \omega_{opt}$ can be interpreted as a numerical

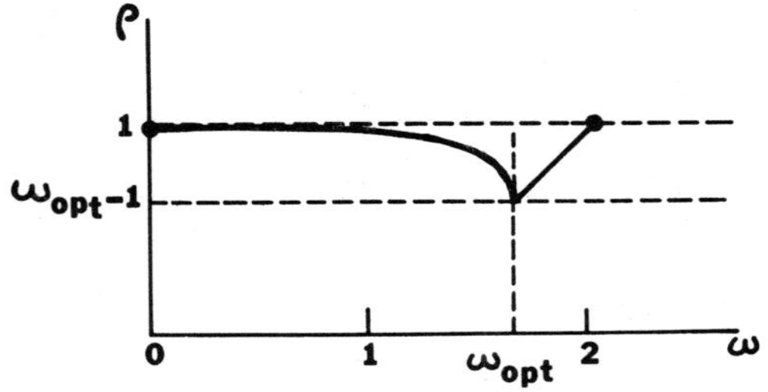

Figure 3. Dependence of spectral radius upon the over-relaxation parameter

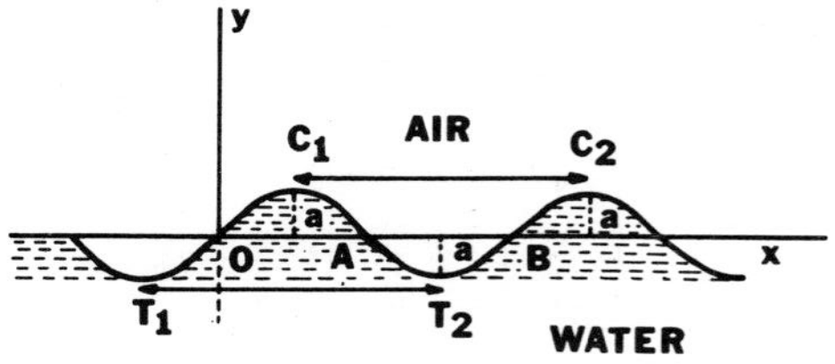

Figure 4. Periodic progressing gravity waves

approximation to the equation

$$2\pi \frac{\partial v}{\partial t} = \nabla^2 v - (v_{xt} + v_{yt}) \; . \tag{3.9}$$

We have thus achieved a very substantial increase in convergence by solving (3.9) and letting time tend to infinity instead of solving (3.3). It is plausible that similar arguments will apply to the solution of FBPS using MBPS.

4. INSTABILITY AND NONUNIQUENESS

In this section we discuss special circumstances under which the solution of a FBP as the limit of a MBP either is or is not appropriate.

Case 1. Unstable FBPS. If a FBP is unstable then it may be difficult or impossible to approximate it as the limit of a MBP. There is considerable evidence that many fluid mechanics FBPS are unstable (Cryer [5, p. 106]). It might be thought that an unstable FBP would present mathematical difficulties, but this is not always the case as is shown by the example of periodic progressing gravity waves, that is, waves on the surface of a liquid such as water which progress with unchanged form (Figure 4). By considering the problem with respect to coordinate axes moving with the waves, we obtain a FBP. The existence of periodic progressing waves was first proved by Nekrassov and Levi-Civita in the 1920's and there are now many existence proofs available (Cryer [6, appendix 2, p. 81]). However, there is both experimental and theoretical evidence that periodic progressing waves are unstable (Benjamin [2], Benjamin and Feir [3], and Roskes [13]).

Case 2. FBPS with non-unique solutions. If a FBP has more than one solution then one possible approach is to solve a corresponding MBP in the hope that the solution of the MBP will tend to a physically meaningful solution.

The fact that certain FBPS have multiple solutions came as rather a surprise to the mathematicians of 60 years ago. The first example was given by Brillouin in a famous paper of 1911, and subsequently many other examples were given, notably by Villat. In all these examples, the authors showed the existence of two or more solutions. However, Zarantonello [15] turned the problem around. Zarantonello considered the two-dimensional irrotational incompressible flow of a stream past a flat plate under the following assumptions: (i) The flow is a uniform parallel flow at infinity; (ii) The velocity is everywhere finite; (iii) On FBS, which are the boundaries of cavities, the normal velocity of the fluid is zero and the tangential velocity is constant. Zarantonello showed that this FBP has eight or nine solutions (see Figure 5, in which cavities are hatched and where solution d may not exist).

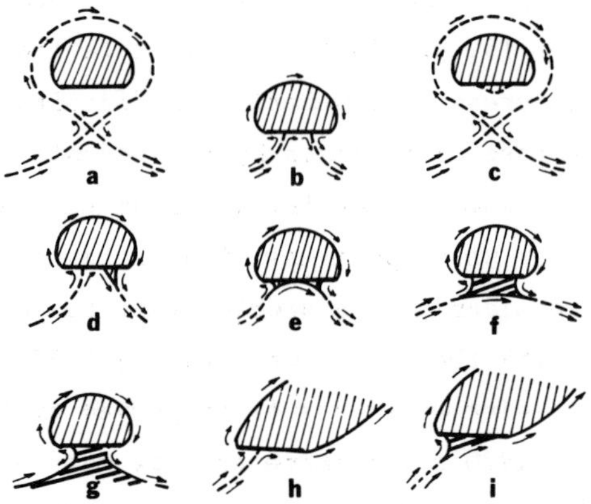

Figure 5. Nine possible parallel cavity flows past a plate

It is often felt that mathematical difficulties arise because the model being used does not adequately represent the physical problem. For example, in the formulation of the problem of flow past a plate shown in Figure 5 important effects such as viscosity have been neglected, and one may reasonably suspect that this is the reason for the non-uniqueness. However, in the remainder of this talk we will give four examples which show that problems with stable non-unique solutions do exist in nature.

Example 1. Elastic-plastic problems. In Figure 6 the stress σ in an elastic-plastic material is shown as a function of the strain ϵ (Hill [11, p. 9]). As long as the stress increases monotonely so does the strain. However, if we are at the point S, and the stress unloads (decreases) to zero, then the strain does not decrease to zero but rather to the positive value represented by O' in Figure 6. On increasing the strain again, the curve O'P'Y' is followed. This non-unique relationship between stress and strain implies that if we take a piece of elastic-plastic material and load it, then the final state may be non-unique because it depends not only upon the final applied stress but also upon the loading history. This is one reason why elastic-plastic FBPS are often solved by the incremental method (Cryer [6, appendix 5, p. 59]).

An example of a problem in which unloading does occur is shown in Figure 7. A hollow square bar is twisted. As the torque increases, regions of plasticity develop and grow until the entire bar is plastic. It has been shown by Herakovich and Hodge [10] that although the applied torque steadily increases, unloading does occur.

Although Example 1 shows the possibility of non-uniqueness, it is perhaps not completely satisfactory because it still requires the use of a model. In the remainder of this lecture I will, therefore, attempt to demonstrate three examples of non-uniqueness to you.

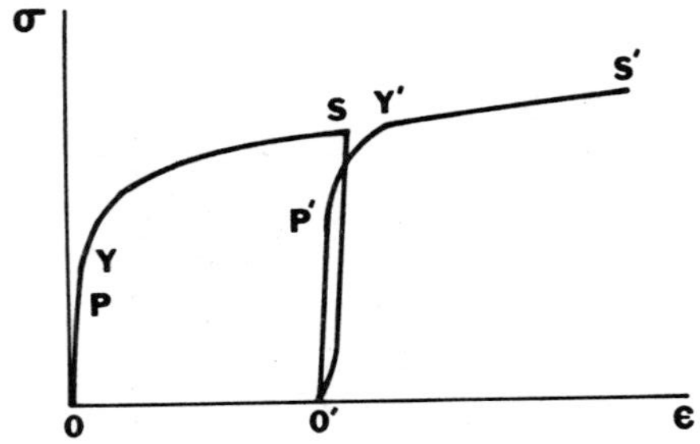

Figure 6. Stress-strain relations for an elastic-plastic material

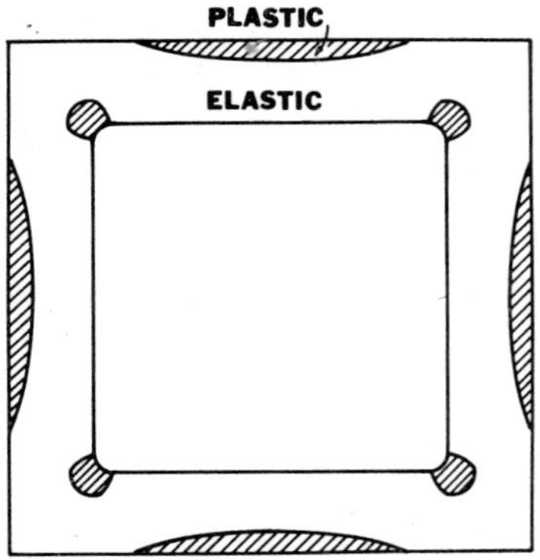

Figure 7. Elastic-plastic torsion of a hollow bar

Example 2. The Borda mouthpiece. A Borda mouthpiece consists of a tube jutting into a container (see Figure 8). If the container is filled with water then the water flows out through the tube. The flow which is usually considered in the literature is the laminar flow shown in Figure 8a, in which a clear laminar jet of water flows out without touching the inside of the tube. A second flow is possible in which the flow is frothy and turbulent (Figure 8b). This second type of flow is sometimes mentioned in the literature (Gibson [9, p. 122]) but the circumstances under which it occurs are not given. In the demonstration apparatus here, the dimensions of the tube were found by trial-and-error; it is 1-1/4" long with internal diameter 5/16" except for a 1/2" section with diameter 1/4" which begins 1/4" from the outer end. The container is 5" wide, 17" long, and 24" high.

When I remove a plug at the end of the tube, the observed flow is turbulent. However, by blowing air down a small pipe into the outer entrance of the tube, the water is blown away from the tube walls and the flow becomes laminar. To obtain the turbulent flow again, I merely obstruct the jet momentarily.

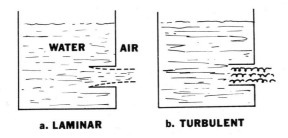

Figure 8. Flows through a borda mouthpiece

Example 3. Channel flow. In this demonstration water flows down a channel over an obstacle (Figure 9). In the demonstration apparatus, the channel is 1" wide, 3" high, and 20" long, and the slope is approximately 1:5. The water is provided by a siphon from the large container. Figure 9 was drawn before the apparatus was built, and in the demonstration the obstacle is more symmetric than shown and the flows are rather different.

Two flows are possible. In the first flow the water flows over the obstacle in a thin layer (Figure 9a); this is called supercritical flow. In the second flow, the water piles up behind the obstacle and flows over the obstacle in a somewhat thicker layer (Figure 9b); this is called subcritical.

To get from supercritical flow (Figure 9a) to subcritical flow (Figure 9b) I merely place my fingers on the obstacle for a few seconds forcing the water to build up behind the obstacle. To get from subcritical flow (Figure 9b) to supercritical flow (Figure 9a), I use my fingers to sweep the excess water downstream.

Example 4. Balloons. A number of years ago I was helping to blow balloons at a birthday party. While engaged in this not too arduous task, it occurred to me that the process of blowing up a balloon is a MBP, the FB being the surface of the balloon.

Most of the balloons were round or cylindrical and caused no difficulties. However, some of the balloons consisted of large cavities connected by narrower channels. When I first blew up one of these balloons, I obtained the shape shown in Figure 10a: the first cavity became large and round, while the remaining cavities remained small. Since this did not look very nice, I let the balloon deflate and adopted a different strategy. I first blew moderately hard, so that the first cavity became a small round ball. I then held the opening closed and squeezed the air from the first cavity along the balloon into the last cavity, where it remained, before blowing in more air. In this way I was able to obtain the desirable shape shown in Figure 10b.

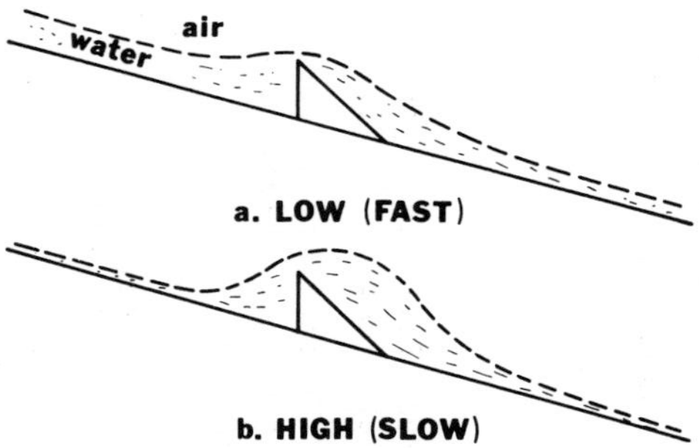

Figure 9. Flows down an inclined plane

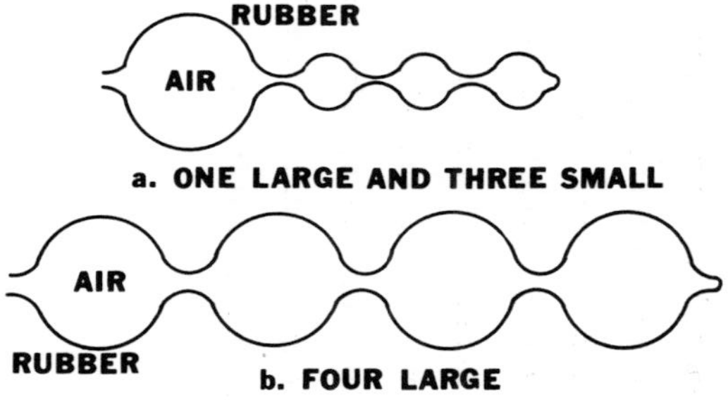

Figure 10. Balloons

Unfortunately, I cannot demonstrate this experiment to you because I have been unable to find the same balloons; I have found balloons with the right shape, but the rubber stretches too easily. I can, however, demonstrate a related experiment which was discovered by my son Martin. I have here a rather tough cylindrical balloon, which I partly blow up and then close. As you see, there is a bubble near the opening with a long tail. I can squeeze the bubble along the balloon and it stays wherever it is put, so that we have a demonstration of a FBP with infinitely many solutions.

ACKNOWLEDGEMENTS

The Borda mouthpiece experiment was demonstrated to me many years ago by Mr. (now Professor) T. B. Benjamin, who kindly provided the approximate dimensions. Professor P. L. Monkmeyer helped in the design of the portable apparatus to demonstrate the Borda mouthpiece and channel flow, and the apparatus was expertly built by Mr. R. C. Hughes.

REFERENCES

1. A. J. Acosta. Hydrofoils and hydrofoil craft, Annual Review of Fluid Mechanics, 5, (1973), pp. 161-184.

2. T. B. Benjamin. Instability of periodic wavetrains in nonlinear dispersive systems, Proc. Roy. Soc. London, 299A, (1967), pp. 59-75.

3. T. B. Benjamin and J. E. Feir. The disintegration of wave trains on deep water. Part I. Theory, J. Fluid Mech., 27, (1967), pp. 417-430.

4. L. Crocco. A suggestion for the numerical solution of the steady Navier-Stokes equations, AIAA (Amer. Inst. Aeron. Astron.) J., 3, (1965), pp. 1824-1832.

5. C. W. Cryer. A survey of trial free-boundary methods for the numerical solution of free boundary problems, Tech. Sum. Rept. No. 1693, Math. Res. Center, University of Wisconsin, November 1976.

6. C. W. Cryer. A bibliography of free boundary problems, Tech. Sum. Rept. No. 1793, Math. Res. Center, University of Wisconsin, October 1977.

7. G. E. Forsythe and W. R. Wasow. Finite Difference Methods for Partial Differential Equations, John Wiley, New York, 1960.

8. P. R. Garabedian. Estimation of the relaxation factor for small mesh size, Math. Tables Aids Comput., 10, (1956), pp. 183-185.

9. A. H. Gibson. Hydraulics and its Applications, Fifth Edition, Constable and Co., London, 1952.

10. C. T. Herakovich and P. G. Hodge, Jr. Elastic-plastic torsion of hollow bars by quadratic programming, Internat. J. Mech. Sci., 11, (1969), pp. 53-63.

11. R. Hill. The Mathematical Theory of Plasticity, Clarendon Press, Oxford, 1950.

12. R. T. Knapp, J. W. Daily and F. G. Hammitt. Cavitation, McGraw-Hill, New York, 1970.

13. G. J. Roskes. Nonlinear multiphase deep-water wave trains, Physics of Fluids, 19, (1976), pp. 1253-1254.

14. R. S. Varga. Matrix Iterative Analysis, Prentice-Hall, Englewood Cliffs, 1962.

15. E. H. Zarantonello. Parallel cavity flows past a plate, J. Math. Pures. Appl., 33, (1954), pp. 29-80.

C. W. Cryer
Mathematics Research Center
University of Wisconsin
610 Walnut Street
Madison, WI 53706

Research sponsored by the National Science Foundation under Grant No. DCR75-03838, and by the United States Army under Contract DAAG29-75-C-0024.

NUMERICAL METHODS FOR ALLOY SOLIDIFICATION PROBLEMS

George J. Fix

Alloy solidification problems differ from classical Stefan problems in that the melting temperature is not known a priori. Because of this numerical techniques and analytical concepts differ from the classical Stefan setting. This paper reviews the fundamental work of Langer and others on a "reduced" model. In addition, an apparently new variational characterization of the complete alloy solidification is given.

1. INTRODUCTION

This paper contains a preliminary investigation of the alloy solidification problem. While this is a free boundary problem it differs significantly from the classical Stefan problem in that the melting temperature is not known a priori; the latter typically depends on the concentration of the "impurities" in the alloy. As a consequence the mathematical theory and the types of numerical methods used in this setting differ rather sharply from the classical context defined by Stefan problems.

In the next section we shall formulate alloy solidification as a free boundary problem. In the third section we show that in certain limiting circumstances, alloy solidification reduces to a "generalized" two phase Stefan problem; i.e., one where the free surface condition involves the curvature of the free surface. The latter can be reformulated as an integral equation, which is useful both for theoretical and computation purposes. We shall

ISBN 0-12-757350-X

review the fundamental work of Langer and others which deal with this model.

The final section contains an apparently new weak variational formulation of the general problem. From this formulation various finite difference approximations are obtained.

2. THE BASIC FREE BOUNDARY PROBLEM

Following Chalmers [1] we consider a binary alloy which occupies the region Ω in space (e.g., a carbon-iron material). We let

$$c = c(x,t)$$

denote the concentration of one of the components of the alloy, and for purposes of exposition we shall call the latter the "impurity" of the alloy. The other variable of interest is temperature

$$u = u(x,t).$$

We are interested in the case where the alloy is initially in the liquid phase and begins to solidify as time increases. We let $\Omega_+(t)$, $\Omega_-(t)$ denote the subregions of Ω occupied by the liquid and solid parts at time t, and let $\mathcal{L}(t)$ denote the interface separating these regions. In each of these regions the temperature and concentration satisfy a Fourier heat law, i.e.,

$$\left.\begin{aligned}\frac{\partial u}{\partial t} &= \nabla(\alpha\nabla u) \qquad (2.1)\\ \frac{\partial c}{\partial t} &= \nabla(\beta\nabla c) \qquad (2.2)\end{aligned}\right\} \text{ for } \underline{x}\in\Omega_{\pm}(t)\,,\ t>0$$

The diffusion constants α, β will typically change discontinuously across the free surface and we denote the restrictions of these functions to the liquid and solid phases by $\alpha_{\pm}$, $\beta_{\pm}$. More generally we shall let $\xi_{\pm}$ denote the restriction of a function ξ defined in Ω to $\Omega_{\pm}(t)$.

The temperature gradient

$$\underline{q} = \alpha\underline{\nabla}u$$

jumps across the free surface due to the liberation of latent heat, and the jump

$$[\alpha\underline{\nabla}u]_{-}^{+} = \alpha\nabla u|_{+} - \alpha\nabla u|_{-}$$

is governed by the classical Stefan condition [2]. Indeed, let $\mathcal{L}(t)$ be described in space-time by

$$F(\underline{x},t) = 0,$$

then

$$\lambda\frac{\partial F}{\partial t} + \underline{\nabla}F\cdot[\alpha\nabla u]_{-}^{+} = 0, \quad \text{for } \underline{x}\in\mathcal{L}(t)\,, \qquad (2.3)$$

where $\lambda > 0$ is the latent heat.

The jump condition for the concentration c across $\mathcal{L}(t)$ is somewhat more elaborate. In particular, thermodynamic considerations [1] yield the eutectic diagram shown in Figure 2.1. The latter should be interpreted as follows. If at any point in the material the concentration of the impurity is c_A and the temperature is u (i.e., point A in Figure 2.1), then this point is in the solid region. On the other hand, if the concentration is raised to a higher value c_B, then the point is in the liquid region. Concentrations in the range

$$c_-(u) < c < c_+(u)$$

are physically unstable, and the only possibilities are for $c \to c_-(u)$ and the point to be in the solid region, or $c \to c_+(u)$ and the point is in the liquid region. In short, the melting temperature at any point is going to depend on the concentrations of the impurity. For simplicity we shall assume that the lines separating the phases are linear, i.e.,

$$c_-(u) = \sigma_- u \, , \quad c_+(u) = \sigma_+ u \tag{2.4}$$

for given constants $\sigma_{\pm}$ (see [1]).

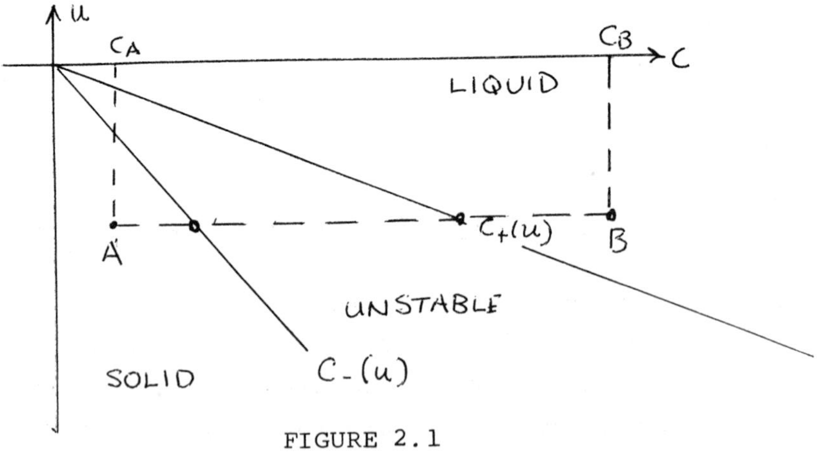

FIGURE 2.1

The above considerations suggest that there is a jump

$$[c]_-^+ = c_+ - c_-$$

in the concentration across the free surface as well as a jump

$$[\beta\underline{\nabla}c]_-^+ = \beta\underline{\nabla}c|_+ - \beta\underline{\nabla}c|_-$$

in the flux. These jumps are governed by a Stefan type condition [1], namely

$$\frac{\partial F}{\partial t}[c]_-^+ + \underline{\nabla}F \cdot [\beta\underline{\nabla}c]_-^+ = 0 \text{ for } \underline{x}\in\mathcal{L}(t). \qquad (2.5)$$

Finally, the concentration and temperature equations are coupled at the free surface by

$$\left.\begin{array}{ll} u_+ = u_- = g & (2.6) \\ c_+ = \sigma_+ g, \ c_- = \sigma_- g & (2.7) \end{array}\right\} \text{for } \underline{x}\in\mathcal{L}(t),$$

where g is the unknown melting temperature.

To complete the description of the problem we must supply initial data

$$u = u_0, \ c = c_0 \text{ on } t = 0 \qquad (2.8)$$

and boundary data. The latter are typically mixtures of Neuman and Dirichlet conditions, say

$$u = u_\Gamma \text{ on } \Gamma_D^{(1)}, \quad \underline{\nabla}u\cdot\underline{\nu} = q_\Gamma^{(1)} \text{ on } \Gamma_N^{(1)} \qquad (2.9)$$

and

$$c = c_\Gamma \text{ on } \Gamma_D^{(2)}, \quad \underline{\nabla} c \cdot \underline{\nu} = \underline{q}_I^{(2)} \text{ on } \Gamma_N^{(2)},$$

where $\Gamma = \Gamma_D^{(i)} \cup \Gamma_N^{(i)}$ $(i = 1,2)$ is the boundary of $\Omega = \Omega_+(t) \cup \Omega_-(t)$ and $\underline{\nu}$ is the outer normal.

The complete system is the two partial differential equations (2.1)-(2.2), the jump conditions (2.3) and (2.5), the interface conditions (2.6)-(2.7), and the initial boundary conditions (2.8)-(2.10). The unknowns are the concentration and temperature fields u and c, the melting temperature g, and the free surface F.

3. INTEGRAL EQUATIONS AND LANGER'S SYMMETRIC MODEL

In this section we present a simplified model due to Langer and others ([2]-[4]). The model is of fundamental importance for the general problem since it contains nonlinear instabilities -- commonly called Mullins-Sekerka instabilities [5] -- associated with the growth of dendrites.

To fix ideas let us consider momentarily the one dimensional case where

$$\Omega = [0,1]. \tag{3.1}$$

Letting

$$x = s(t) \tag{3.2}$$

denote the position of the free surface, so

$$\Omega_- = (0, s(t)), \quad \Omega_+ = (s(t), 1), \tag{3.3}$$

the equation for the temperature field is

$$\frac{\partial u}{\partial t} = \frac{\partial}{\partial x}\alpha\frac{\partial u}{\partial x} \text{ in } \Omega_{\pm}(t), \quad t > 0. \tag{3.4}$$

The jump condition reduces to

$$\lambda\frac{ds}{dt} + [\alpha\frac{\partial u}{\partial x}]_{-}^{+} = 0 \quad \text{for } x = s(t) \tag{3.5}$$

with say $u = u_{\Gamma}^{(0)}, u_{\Gamma}^{(1)}$ for $x = 0,1$.

The first simplification is to let $\alpha \uparrow \infty$. This implies that the temperature field is given by

$$u(x,t) = \begin{cases} [u_{\Gamma}^{(1)} - u_{\Gamma}^{(0)}]x + u_{\Gamma}^{(0)} & \text{for } 0 \leq x \leq s(t) \\ [u_{\Gamma}^{(1)} - u_{\Gamma}^{(0)}](x-1) + u_{\Gamma}^{(1)} & \text{for } s(t) \leq x \leq 1 \end{cases} \tag{3.6}$$

In particular the melting temperature is

$$g(s(t)) = [u_{\Gamma}^{(1)} - u_{\Gamma}^{(0)}]s(t) + u_{\Gamma}^{(0)} \tag{3.7}$$

The equations governing the concentration are

$$\frac{\partial c}{\partial t} = \frac{\partial}{\partial x}\beta\frac{\partial c}{\partial x} \text{ in } \Omega_{\pm}(t), \ t > 0, \tag{3.8}$$

and

$$\frac{ds}{dt}[c]_{-}^{+} + [\beta\frac{\partial c}{\partial x}]_{-}^{+} = 0 \text{ for } x = s(t) \tag{3.9}$$

plus initial and boundary conditions. The Langer

model is completed by making two final assumptions. The first is the assumption that β is constant in Ω, and the second is that $[c]_-^+$ is constant. The latter assumption is equivalent to replacing the eutectic diagram in Figure 2.1 with the one shown in Figure 3.1.

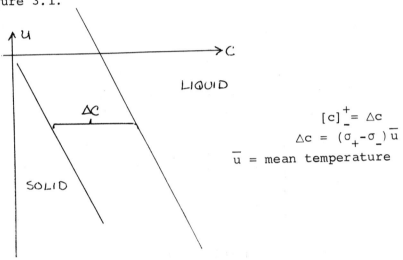

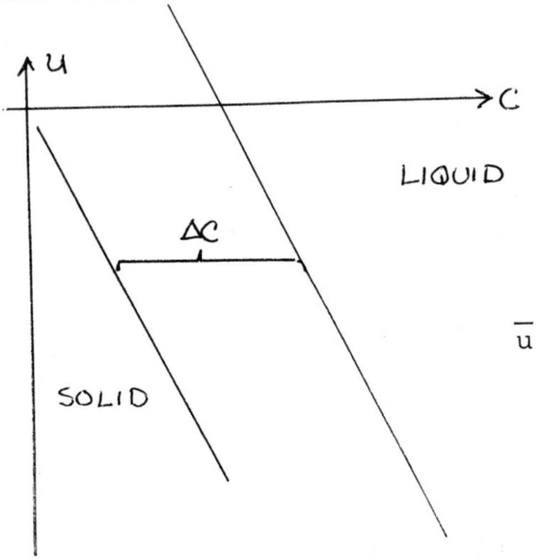

$$[c]_-^+ = \Delta c$$
$$\Delta c = (\sigma_+ - \sigma_-)\,\bar{u}$$
$$\bar{u} = \text{mean temperature}$$

FIGURE 3.1

Letting c_o be the equilibrium concentration defined by

$$c_o(x,t) = \begin{cases} 0 & \text{if } x < s(t) \\ \Delta c & \text{if } x \geqq s(t)\,, \end{cases} \tag{3.10}$$

and letting

$$c(x,t) = c_o(x,t) + \frac{w(x,t)}{\Delta c} \tag{3.11}$$

we have the following generalized two-phase Stefan problem for w:

$$\frac{\partial w}{\partial t} = \frac{\partial}{\partial x}\left(\beta \frac{\partial w}{\partial x}\right) \text{ in } \Omega_{\pm}(t), \; t > 0 \tag{3.12}$$

with

$$\left.\begin{aligned} &\frac{ds}{dt} + [\beta \nabla w]_{-}^{+} = 0 \\ &w(s(t),t) = \sigma_{-} g(s(t)) \end{aligned}\right\} \text{ for } x = s(t) \tag{3.13}$$

plus initial and boundary conditions on t = 0, x = 0, and x = 1.

The two and three dimensional analogs of (3.12)-(3.13) are derived in an analogous manner. In particular suppose the free surface is defined by

$$z = \zeta(\underline{x},t) \tag{3.14}$$

as shown in Figure 3.2. Then

$$\frac{\partial w}{\partial t} = \nabla(\beta \nabla w) \text{ in } \Omega_{\pm}(t), \; t > 0 \tag{3.15}$$

and the interface conditions are

$$\frac{\partial \zeta}{\partial t} + [\beta \nabla w]_{-}^{+} = 0 \text{ for } z = \zeta(x,t) \tag{3.16}$$

$$\frac{w}{\Delta c} = \xi \; \kappa\{\zeta\} \tag{3.17}$$

where κ is the curvature of the free surface and ξ a capillary length.

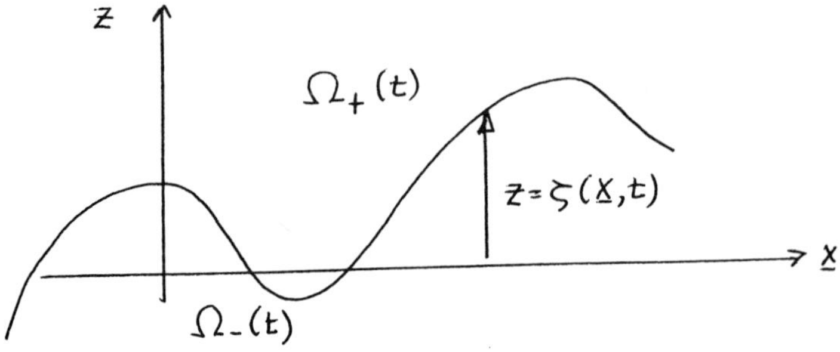

FIGURE 3.2

It is precisely in the multidimensional context that the Mullins-Sekerka instabilities can occur. In the latter one has initially a planar interface, say $\zeta = 0$. In the presence of thermal gradients this interface will deform revealing a wealth of structure including multiple bifurication points with different stability properties. The latter suggests at least qualitatively a dendritic breakaway mechanism.

To review the essential features of Langer's results we shall assume Ω extends from $z = -\infty$ to $z = \infty$ and use the boundary condition

$$\lim_{z\to\pm\infty} \frac{dw}{dz} = \left(\frac{dw}{dz}\right)_\infty .$$

Define the Green's function by

$$G(z,\underline{x},t|z_1,\underline{x}_1,t) = \theta(t-t_1)\exp[-R]/[4\pi\beta(t-t_1)]^{3/2}, \tag{3.18}$$

where $R = [|z-z_1|^2+|\underline{x}-\underline{x}_1|^2]/[4\beta(t-t_1)]$ and where θ is unity when its argument is positive and vanishes otherwise. Then the following integral equation for the free surface holds:

$$\xi\kappa\{\zeta(\underline{x},t\} + \frac{1}{\ell}\zeta(\underline{x},t) = \int_{-\infty}^{t} dt_1 \int d\underline{x}_1 G(\underline{x},\zeta,t|\underline{x}_1,\zeta_1,t_1)\,(d\zeta_1/dt_1) \quad (3.19)$$

where $\zeta = \zeta(\underline{x},t)$, $\zeta_1 = \zeta(\underline{x}_1,t_1)$, and ℓ is the diffusion length

$$\frac{1}{\ell} = -\frac{1}{\Delta c}\left(\frac{dw}{dz}\right)_{\infty}$$

(see [2]).

Observe that $\zeta = 0$ is always a solution of (3.19). To study the stability of this solution Langer and Turski [2] linearized (3.19) about this solution and looked for solutions in the form

$$\zeta(\underline{x},t) = \Sigma_{\underline{k}} A_{\underline{k}}(t) \exp(i\underline{k}\cdot\underline{x})$$

The resulting equations for the coefficients $A_{\underline{k}}$ were then analyzed with standard bifurication techniques. A wealth of branching processes were obtained that are too diverse to report here. While the details of this analysis may be in question because of the linearizations, it does show that one must be prepared to expect nonuniqueness through branching and in fact unstable branches in alloy solidification.

The linearized stability analysis is of fundamental importance for numerical simulations used in conjunction with the integral equation (3.19). The idea is to use the linearized analysis to identify branch points and then use projective techniques on (3.19) to continue the branches. Results based on these techniques will be reported in a subsequent publication.

4. A WEAK VARIATIONAL FORMULATION

The simplified model discussed in the previous section is of crucial importance because of the insight it offers. However, the limitations of the model are obvious, not only because of the restrictive assumptions involved, but also because of its dependence on the solution of a complicated nonlinear integral equation. The latter are notorious for numerical instabilities, and preliminary work with (3.19) indicates that it may suffer some of these defects. In this section we develop an alternative to (3.19) based on a weak variational formulation of the complete problem defined in Section 2.

If we assume for the moment that the melting temperature g is known, then the temperature and concentration equations decouple. A weak formulation for the former in this context is simply a description of heat balance in terms of the heat function

$$H(u) = \begin{cases} u-\lambda & \text{if } u-g < 0 \\ u & \text{if } u-g > 0. \end{cases} \tag{4.1}$$

The constant $\lambda > 0$ is the latent heat. In particular, we have

$$\int_{\mathcal{S}} \{H(u)\, n_t + \alpha \underline{\nabla} u \cdot \underline{n}_x\} dS = 0, \tag{4.2}$$

where $\mathcal{S}$ is any closed surface in space-time and $\underline{n} = (\underline{n}_x, n_t)$ is the outer normal to $\mathcal{S}$ with n_t being the time component and $\underline{n}_x$ being the spatial component. The diffusion equation (2.1) can be formally derived from (4.2) by integration by parts and letting $\mathcal{S}$ shrink to a point $(\underline{x},t)$ in space-time (where $\underline{x} \notin \mathcal{L}(t)$). Similarly, the jump condition (2.3), which can be rewritten as

$$[H]_-^+ \frac{\partial F}{\partial t} + \underline{\nabla}F \cdot [\alpha \underline{\nabla} u]_-^+ = 0,$$

is a consequence of (4.2).

Meyer and others ([5]-[7]) have developed schemes based on (4.2). To fix ideas consider the case of one space dimension where (4.2) reduces to

$$\int_{\mathcal{C}} \{H dx + \alpha \frac{\partial u}{\partial x} dt\} = 0, \tag{4.3}$$

for any closed curve $\mathcal{C}$ is space-time. Let us introduce a grid

$$x_o < x_1 < \ldots < x_{N+1}$$

is space and a grid

$$0 = t_o < t_1 < \ldots < t_M$$

and a grid in time. For simplicity let us take these grids to be uniform with

$$\Delta x = x_{j+1} - x_j, \ \Delta t = t_{k+1} - t_k.$$

Consider the closed contour $\mathcal{C}$ shown in Figure 4.1.

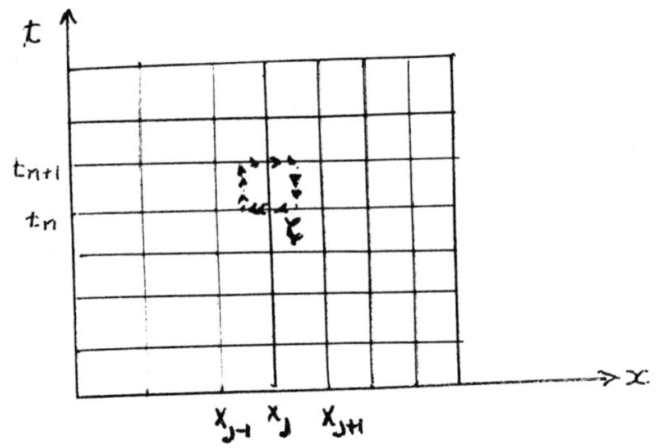

FIGURE 4.1

If we let

$$u_j^n = u(x_j, t_n), \quad H_j^n = H(u_j^n),$$

then a second order approximation to (4.3) is

$$\Delta x\{H_j^{n+1} - H_j^n\} - \Delta t\{(\alpha \frac{\partial u}{\partial x})_{j+1/2}^{n+1/2} - (\alpha \frac{\partial u}{\partial x})_{j-1/2}^{n+1/2}\} = 0, \quad (4.4)$$

where

$$\xi^{n+1/2} = (\xi^n + \xi^{n+1}), \quad \xi_{j+1/2} = (\xi_j + \xi_{j+1})/2.$$

Replacing the derivatives with second order difference quotients we obtain Meyer's second order scheme

$$(H_j^{n+1} - H_j^n)/\Delta t = \{\alpha_{j+1}^{n+1/2} - u_j^{n+1/2}) - \quad (4.5)$$
$$-\alpha_{j-1/2}^{n+1/2}(u_j^{n+1/2} - u_{j-1}^{n+1/2})\}/\Delta x^2.$$

If we knew the melting temperature g, (4.1) and (4.5) plus the boundary and initial conditions would yield the temperature field u^n.

To compute g we must of course bring in the concentration equation. There appears to be some confusion in the literature concerning the correct weak formulation for the latter. One is tempted to use

$$\int_{\mathcal{L}} \{ c n_t + \beta \underline{\nabla} c \cdot \underline{n}_x \} dS = 0 \tag{4.6}$$

because both the partial differential equation (2.2) and the jump condition (2.5) can be formally derived from (4.6). However, difference schemes based on (4.6) are typically divergent. To be precise, in one spatial dimension (4.6) reduces to

$$\int_{C} \{ c dx + \beta \frac{\partial c}{\partial x} dt \} = 0. \tag{4.7}$$

Using the arguments leading to (4.5) we have

$$\begin{aligned}(c_j^{n+1} - c_j^n)/\Delta t = \{ &\beta_{j+1/2}^{n+1/2}(c_{j+1}^{n+1/2} - c_j^{n+1/2}) - \\ &- \beta_{j+1/2}^{n+1/2}(c_j^{n+1/2} - c_{j-1}^{n+1/2}) \}/\Delta x^2 .\end{aligned} \tag{4.8}$$

If the concentration field $\{c_j^n\}$ is known, then (4.8) along with the boundary conditions determine $\{c_j^{n+1}\}$. Knowing this we can determine the free surface -- the jump condition (2.5) is implicit in (4.8) -- and hence the melting temperature. The relation (4.5) can then be used to determine the temperature field u. Unfortunately, these approximations are typically divergent even for very simple situations.

Numerical results showing divergence will be reported in a subsequent publication.

The problem is that while (4.6) is in some sense a weak formulation, it cannot be interpreted literally. (This was done in the derivation of (4.8)). The concentration c is discontinuous across the free surface and hence the term

$$\int_{\mathcal{S}} \beta \underline{\nabla} c \cdot \underline{n}_x \, dS$$

must be interpreted as a distribution.

To derive a more suitable weak formulation, let us first take $\mathcal{S}$ to be a closed surface in space-time which includes part of the free surface as in Figure 4.2. From the partial differential equations we know

$$0 = \int_{\mathcal{R}_{\pm}} \{\frac{\partial c}{\partial t} - \underline{\nabla} \beta \underline{\nabla} c\} d\underline{x}\, dt$$

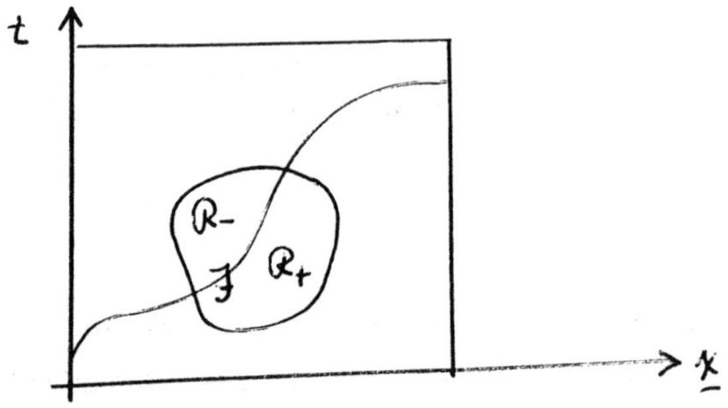

FIGURE 4.2

After integrating we obtain

$$\int_{\partial\mathcal{R}_{\pm}} \{ c n_t^{\pm} + \beta\underline{\nabla}c \cdot n_{\underline{x}}^{\pm} \} ds = 0, \tag{4.9}$$

where $(n_{\underline{x}}^{\pm}, n_t^{\pm})$ is the outer normal to $\mathcal{R}_{\underline{+}}$. The sum of these two integrals leads first of all to an integral

$$\int_{\mathfrak{F}} \{ [c]_-^+ n_t^+ + [\beta\underline{\nabla}c]_-^+ \cdot \underline{n}_{\underline{x}}^+ \} ds \tag{4.10}$$

over the portion $\mathfrak{F}$ of the free surface. This contribution to the sum is zero by the jump conditions, hence we are left with

$$\left\{ \int_{\mathcal{S}_+} + \int_{\mathcal{S}_-} \right\} \{ c n_t + \beta\underline{\nabla}c \cdot \underline{n}_{\underline{x}} \} ds = 0 \tag{4.11}$$

Observe that (4.6) is obtained from (4.11) by replacing the sum of integrals over $\mathcal{S}_+$ and $\mathcal{S}_-$ with a single integral over $\mathcal{S}$. This, however, is the problem with (4.6). Since c is discontinuous across the free surface, the left hand sides of (4.6) and (4.11) are not equal, i.e., (4.6) is false if taken in a literal sense.

To express (4.11) in a more convenient form for deriving difference equations, we recall that at the free surface

$$c_+ = \sigma_+ g, \quad c_- = \sigma_- g;$$

Letting σ be the piecewise constant function defined by

$$\sigma = \begin{cases} \sigma_+ & \text{in the liquid region} \\ \sigma_- & \text{in the solid region,} \end{cases}$$

we see that

$$c/\sigma$$

is continuous across the free surface and in fact is equal to the melting temperature there. This suggests we define g in all of space-time by

$$g = c/\sigma. \tag{4.12}$$

Then (4.11) can be written

$$\int_{\mathcal{S}} \{cn_t + (\beta\sigma)\,\underline{\nabla}g\cdot\underline{n}_x\}ds = 0 \tag{4.13}$$

We summarize the total formulation in the following.

Weak variational principle for alloy diffusion. Find a pair of functions c, u such that the following are valid:

(a) c is square integrable and $g = c/\sigma$ has a square integrable gradient

(b) (4.13) is valid for all closed surfaces $\mathcal{S}$

(c) u has a square integrable gradient

(d) (4.2) is valid for all closed surfaces $\mathcal{S}$.

One can derive difference schemes directly from (4.13). For example, in one spatial dimension one has

$$(c_j^{n+1}-c_j^n)/\Delta t = \{\gamma_{j+1/2}^{n+1/2}[g_{j+1}^{n+1/2}-g_j^{n+1/2}] - $$

$$-\gamma_{j-1/2}^{n+1/2}[g_j^{n+1/2}-g_{j-1}^{n+1/2}]\} \quad (4.14)$$

where

$$\gamma = \beta\sigma \quad (4.15)$$

This is a nonlinear set of equations for c^{n+1} (in terms of c^n) since both β and σ (and hence γ) will have different values in the liquid and solid regions. The same is true for (4.5), and following Meyer [6] this set of nonlinear equations is best treated by successive overrelation. Computations with these approximations will be reported in a subsequent publication.

Ockendon [9] has derived a similar weak formulation for the alloy diffusion problem in one spatial dimension. However, the latter is expressed as a partial differential equation $\frac{\partial c}{\partial t} = (\partial^2 q/\partial x^2)$ instead of integral as was done in this paper. These approaches do differ in that the Ockendon's formulations requires that the diffusion coefficient β to be constant in the liquid and in the solid phases. Such a restriction is not needed in the integral formulation.

There are many issues involved with the weak formulation (a)-(d) which we hope to address in future publications. First of all, for those problems where the solution is known to exist and be unique -- i.e., before the onslaught of dendritic instability-- what sort of accuracy can be expected from the difference approximations? Finally, there is the crucial question of how branching processes are to be treated with this type of approximation.

REFERENCES

1. B. Chalmers. Principles of Solidification, John Wiley, New York, 1964.

2. J. S. Langer and L. A. Turski. "Studies in the theory of interface stability, I Stationary symmetric model, to appear in Acta. Met.

3. J. S. Langer and L. A. Turski. "Studies in the theory of interface stability, II Moving symmetric model, to appear in Acta. Met.

4. J. S. Langer and Müller-Krumbhaar. "Theory of dendritic growth", Carnegie-Mellon Research Report, Center for Joining of Materials, 1977.

5. W. W. Mullins and J. Sekerka. J. Appl. Phys., 34, 323 (1963), 35, 444 (1964).

6. G. H. Meyer. SIAM J. Numer. Anal., 10, 522-538 (1973).

7. D. R. Atthey. in Moving Boundary Problems in Heat Flow and Diffusion, ed., J. R. Ockendon and W. R. Hodgkins, Oxford, 1974.

8. A. B. Tayler. in Moving Boundary Problems in Heat Flow and Diffusion, ed., J. R. Ockendon and W. R. Hodgkins, Oxford, 1974.

9. J. R. Ockendon. "Numerical and analytical solutions of moving boundary problems", this volume.

George J. Fix
Department of Mathematics
Carnegie-Mellon University
Pittsburgh, PA 15213

This work was supported in part by the Office of Army Research under Contract No. DAAG29-77-G-0026.

NUMERICAL AND ANALYTIC SOLUTIONS OF MOVING BOUNDARY PROBLEMS

J. R. Ockendon

Numerical schemes are presented for finding unknown boundaries in problems of frictional oscillations and alloy solidification. A quantitative description is given of singularities which arise in moving boundary problems for vapour diffusion in a porous medium and for suction from a Hele-Shaw cell. All these situations emphasise the importance of weak solutions to moving boundary problems.

1. INTRODUCTION

This paper describes four unknown boundary problems arising in industrial situations. The first two are studied with an eye to finding efficient numerical procedures; the others concern the description of the solution when the unknown boundary has a singularity.

If there is an underlying theme to emerge from these studies, it is the theoretical and practical importance of weak formulations of unknown boundary problems. In a very simple case, let $u(x,t)$ be the temperature in a Stefan problem when $x > s(t)$ is liquid and $x < s(t)$ is solid. Then, if L is the latent heat, u not only satisfies the classical formulation

$$u_t = (\beta(u)\ u_x)_x \tag{1.1a}$$

$$\Big[\beta(u)\ u_x\Big]_{x=s-o}^{x=s+o} = -L\ \frac{ds}{dt}\ ,\quad u(s(t),t) = 0 \tag{1.1b}$$

ISBN 0-12-757350-X

together with suitable initial and boundary conditions in $0<t<T$, $0<x<X$, but u is also a weak solution in that there exists a second function $H(x,t)$ such that

$$u = \begin{cases} H & H < 0 \\ 0 & 0 < H < L \\ H-L & L < H \end{cases} \qquad (1.2a)$$

and that $\int_0^T \int_0^X (\phi_t H + \phi_{xx} v)\,dx\,dt =$
(initial and boundary terms) (1.2b)

for all suitable test functions ϕ [10]. Here $v = \int \beta(u)\,du$.

The two great advantages of (1.2) over (1.1) are the relative ease with which the weak solution can be shown to be unique and the fact that suitable finite difference approximations to the "enthalpy formulation" of (1.1), namely

$$H_t = v_{xx}, \quad v = v(H) \qquad (1.3)$$

do converge to this unique weak solution. Such discretisation may be explicit or implicit in time. The Rankine-Hugoniot conditions corresponding to the conservation law formulation (1.3) are just (1.1b) and $\left[u\right]_{x=s-0}^{x=s+0} = 0$.

(1.2) has the further advantage that physically sensible weak solutions may exist when classical ones do not. This occurs, for example, when there is a volume heat source term in (1.1a), in which case (1.1b) no longer necessarily applies [2]. However, weak formulations are not sufficiently well understood to be able to cope with all situations where the classical solution may break down. Indeed, the second and fourth problems to be discussed here fall into this category.

2. FRICTIONAL OSCILLATIONS.

There are many oscillations whose behaviour depends discontinuously on certain velocity components in the system [1]. Hydraulic linkages in aircraft control systems provided our motivation for considering such problems. The simplest model which displays the difficulties encountered with traditional integration packages is

$$\frac{d^2x}{dt^2} + \operatorname{sgn} \frac{dx}{dt} + x = \sin wt, \quad \frac{dx}{dt} \neq 0. \tag{2.1}$$

The second term on the left-hand side expresses the fact that friction opposes the motion. It has not yet been defined at $\frac{dx}{dt} = 0$ and one approach is to return to the physical situation and subdivide the motion into intervals according to whether $\frac{dx}{dt}$ vanishes; there are then unknown boundaries at the ends of these intervals.

An alternative and numerically more attractive approach has been suggested by Taubert [19] who, following an idea of Filippov [6], shows that $\operatorname{sgn} \frac{dx}{dt}$ can be allowed to take arbitrary numerical values between ± 1 at $\frac{dx}{dt} = 0$.

The implicit discretisation of (1.3) suggests another successful numerical procedure. Write (2.1) as

$$\frac{d^2x}{dt^2} - c\frac{dx}{dt} + H + x = \sin wt \tag{2.2a}$$

where c is an arbitrary positive number and

$$\frac{dx}{dt} = U(H) = \begin{cases} (H-1)/c & H>1 \\ 0 & 1>H>-1 \\ (H+1)/c & -1>H\ . \end{cases} \tag{2.2b}$$

Then the discretisation

$$(x_{n+1}-2x_n+x_{n-1})/\delta t^2 - c(x_{n+1}-x_n)/\delta t + H_n + x_n = \sin wt_n \tag{2.3a}$$

$$x_{n+1} - x_n = \delta t\ U(H_n) \tag{2.3b}$$

converges to the unique pair of functions x(t), H(t) which satisfy

$$\int_o^T \{-\left[\frac{d^2\phi}{dt^2} + c\frac{d\phi}{dt} + \phi\right]\frac{dx}{dt} + \phi(H-\sin wt)\}dt = 0. \tag{2.4}$$

Here we have taken $x(o) = \frac{d\mathbf{x}(o)}{dt} = 0$, $T > 0$ is arbitrary and the test functions ϕ are twice differentiable with $\phi(T) = \frac{d\phi}{dt}(T) = 0$. The Rankine-Hugoniot conditions for (2.4) are that x and dx/dt are continuous everywhere.

Since H is a monotonic increasing function of dx/dt, the proof of these statements [15] closely follows that of Oleinik [10] for the Stefan problem. A typical response for a fairly large value of w^{-1} is Fig. 1.

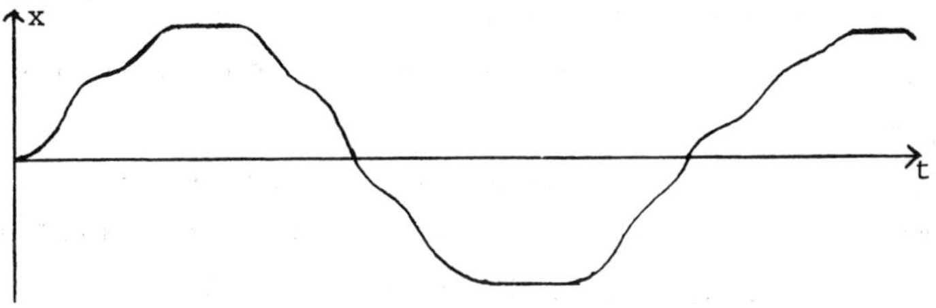

Fig. 1 Solution of $\frac{d^2x}{dt^2} + 0.1\ \mathrm{sgn}\frac{dx}{dt} + x = \sin t/10$ with $x(0) = \frac{dx}{dt}(0) = 0$.

3. ALLOY SOLIDIFICATION

It has been pointed out [18] that the weak formulation of Stefan problems in the form (1.2) does not easily generalise to situations where either the latent heat or the phase change temperature is a function of time. Such effects can occur, even in a simple model for dilute alloy solidification which neglects any coupling between heat and mass transfer effects away from the moving boundary:

$$u_t = \beta u_{xx}, \qquad c_t = \gamma c_{xx} \tag{3.1a,b}$$

$$\left[\beta u_x\right]_{s-0}^{s+0} = -L\frac{ds}{dt}, \qquad \left[\gamma c_x\right]_{s-0}^{s+0} = -\left[c\right]_{s-0}^{s+0}\frac{ds}{dt} \tag{3.2a,b}$$

$$\lim_{x\to s\pm 0} u = u_M(t) \qquad \lim_{x\to s+0} c = c_L\ u_M \tag{3.3a,b}$$

$$\lim_{x\to s-0} c = c_S\ u_M.$$

Here temperatures are denoted by u, concentrations by c, c_S and c_L are prescribed negative constants with $|c_S|<|c_L|$ and β, γ are assumed constant in each phase. (3.2b) expresses conservation of mass at the phase boundary and (3.3) states that the alloy is in equilibrium there (see Tayler [18], who also describes limiting situations in which the solution of (3.1)-(3.3) reduces to (1.1) with variable latent heat and melting temperature).

The first step in writing (3.1)-(3.3) as a generalisation of (1.3) is to note that in the simplest thermodynamic model, (3.1b) is a consequence of

$$c_t = \lambda_{xx} \tag{3.4}$$

where the chemical activity λ is continuous across the phase boundary and is such that $\lambda = \gamma c$ away from this boundary. The Rankine-Hugoniot relations corresponding to this conservation law are

$$\left[\lambda\right]_{s-o}^{s+o} = 0, \quad \left[\lambda_x\right]_{s-o}^{s+o} = -\left[c\right]_{s-o}^{s+o} \frac{ds}{dt} \quad . \tag{3.5}$$

Now (3.4), together with a definition of $c=c(\lambda)$ as a function with a prescribed jump discontinuity at a certain value of λ, would be the analogue of (1.3) for a problem just involving mass transfer. For our problem, we have to define both c and the enthalpy H as functions of u and λ to yield (3.2). To this end we write λ as $-c/c_L$ and $-c/c_S$ in the liquid and solid phases respectively (see[8]). Then, at the phase boundary, $u + \lambda = 0$ so our prescription for c is

$$c = \begin{cases} -c_L\lambda & u + \lambda < 0 \\ -c_S\lambda & u + \lambda > 0. \end{cases} \tag{3.6}$$

The only generalisation of (1.2a) which will conform with (3.2a) is

$$H = \begin{cases} u & u + \lambda < 0 \\ u + L & u + \lambda > 0. \end{cases} \tag{3.7}$$

Inverting (3.6) and (3.7) yields

$$(u,\lambda) = \begin{cases} (H, -c/c_L), & H - c/c_L < 0 \\ (H-L, -c/c_S), & H-L - c/c_S > 0. \end{cases} \tag{3.8}$$

At the phase boundary $u + \lambda = 0$ we add the requirement that u and λ remain independent of time, at their values at which $u+\lambda = 0+$, until H has increased by L and hence c by $(c_L - c_S)u$.

We are thus led to the coupled conservation laws

$$H_t = \beta u_{xx}, \qquad c_t = \lambda_{xx}, \tag{3.9}$$

the solutions of which automatically satisfy (3.2a,b). However our formulation only applies when the thermal conductivity β is the same constant in each phase.

Some results of the explicit discretisation

$$(H, c)_{n+1} - (H, c)_n = \delta t(\beta u_{xx}, \lambda_{xx})_n \tag{3.10}$$

with u and μ defined as above are shown in Fig. 2. There is qualitative agreement with typical temperature and concentration profiles found in zone refinement [4].

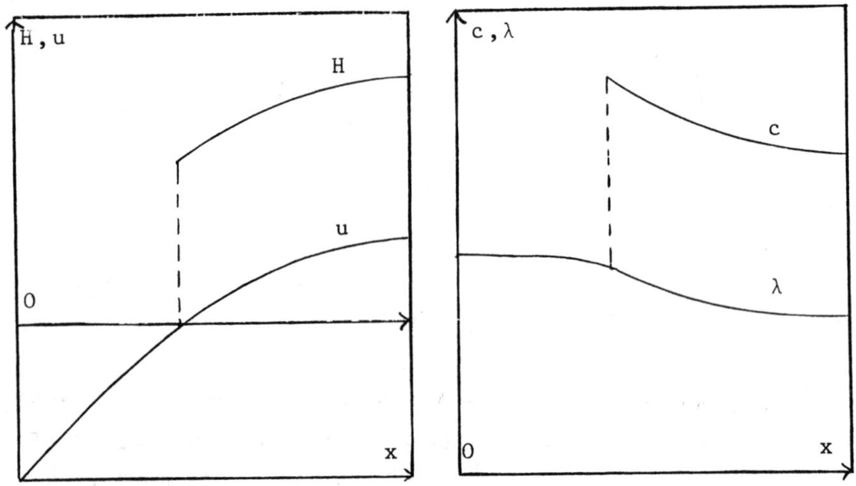

Fig. 2. Solution at $t = 0.25$ of (3.9) with $\beta=1$, $L=1$, $c_L=1$, $c_S=0.5$; $c_x=0$ at $x=0,1$; $u_x=0$ at $x=1$; $u=-1$ at $x=0$; $c=0.1$, $u=1$ at $t=0$.

4. VAPOUR DIFFUSION IN A POROUS MEDIUM

The problem of pollution of a dry porous material by diffusion of water vapour from a confined moist area can be modelled by the degenerate diffusion equation

$$c_t = (cc_x)_x, \quad c(x,0) = c_0(x), \quad -\infty < x < \infty. \qquad (4.1)$$

For a review of the theory of diffusion equations of this type, see Knerr [9]. (4.1) admits several types of similarity solution for suitable initial data; there is also a theory of weak solutions to which certain finite difference approximations converge [7]. These weak solutions may involve a moving boundary $x = s(t)$ at which $c = 0$ and, as is shown in [9],

$$c_x = \frac{ds}{dt} . \qquad (4.2)$$

It is found experimentally that moist regions in otherwise dry porous media can remain confined for periods of days before suddenly expanding on a time scale of hours. This behaviour may be related to the results of Oleinik et al [11], which have been illustrated numerically in [7]. In the simplest case when c_o has compact support and zero gradient at the ends of this support, the vapour does not begin to diffuse outwards until a certain non-zero time interval has elapsed. This phenomenon is related to the existence of the exact solution

$$c = \begin{cases} x^2/1-6t & x < 0 \\ 0 & x > 0 \end{cases} \qquad (4.3a)$$

when $$c_o = \begin{cases} x^2 & x < 0 \\ 0 & x > 0. \end{cases} \qquad (4.3b)$$

We note that c_o does not have compact support, that the solution only exists for $t < 1/6$ and that the boundary $c = 0$ remains fixed throughout that time. The breakdown of (4.3a) at $t=1/6$ precludes its use in predicting any motion of the boundary $c = 0$, and we now present another exact solution which indicates how this boundary begins to move for a certain class of $c_o(x)$.

We consider the case

$$c_o = \begin{cases} x^2-6^{10/3} A\, x^3 + d_o(x) & x < 0 \qquad (4.4a) \\ 0 & x > 0 \qquad (4.4b) \end{cases}$$

where A is arbitrary and $d_o(x)$ has a series representation in ascending powers of x and $\log x$, with coefficients just depending on A in a way to be prescribed below. Also $d_o \sim x^4$ as $x \to 0$. We write, in $t < 1/6$,

$$c = \begin{cases} (t-1/6)^{11/3}\, f\{x/(t-1/6)^{7/3}\} & x<0 \\ 0 & x>0 \end{cases} \tag{4.5}$$

to obtain

$$f\frac{d^2 f}{d\zeta^2} + \left(\frac{df}{d\zeta}\right)^2 + \frac{7}{3}\zeta\frac{df}{d\zeta} - \frac{11f}{3} = 0, \quad \zeta = \frac{x}{(t-1/6)^{7/3}} \tag{4.6a}$$

where $f = -\zeta^2/6 + A\zeta^3 + \tilde{d}_o(\zeta),\ \zeta > 0$ (4.6b)

and $-6^{-11/3}\,\tilde{d}_o\,(-6^{-7/3}x) = d_o(x)$.

Our hope is first to construct d_o so that (4.6b) is indeed a solution of (4.6a) for all $\zeta > 0$; then to find the asymptotic form of (4.6b) as $\zeta \to +\infty$, i.e. $t \to 1/6-0$; and finally, as t increases from 1/6 and ζ increases from $-\infty$, to construct f, still satisfying (4.6a) in such a way that c is continuous at $t = 1/6$. If this last construction means that $f \to 0$ as $\zeta \to \zeta_o > 0$, the solution in $t > 1/6$ is

$$c = \begin{cases} (t - 1/6)^{11/3} f(\zeta) & -\infty<\zeta<\zeta_o \\ 0 & \zeta_o < \zeta \end{cases} \tag{4.7}$$

and the boundary $c = 0$ will move along $x = \zeta_o(t-1/6)^{7/3}$.

We begin by noting that if

$$f = \zeta^2 g, \qquad \frac{df}{d\zeta} = \zeta\, h \tag{4.8a}$$

then $$\frac{dh}{dg} = \frac{11g - 7h - 3h^2 - 3gh}{3g(h-2g)} \quad . \qquad (4.8b)$$

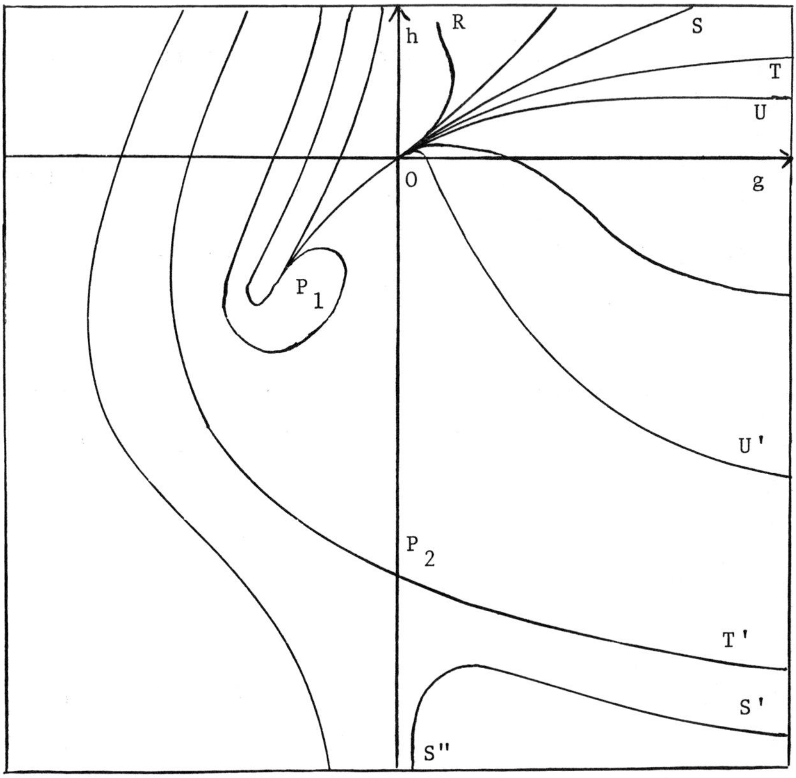

Fig.3 Schematic phase plane for (4.8b)

The phase plane for this equation is shown in Fig.3, there being a degenerate node at the origin 0; numerical results have been used to sketch the unique solution emerging from 0 into the third quadrant. This solution is the only one which permits ζ to increase from 0+ at the node P_1 to $+\infty$ at 0. Now on any phase curve near P_1,

$$h-2g \sim (g+1/6) - \frac{45}{2}(g+1/6)^2 + \ldots + a(g+1/6)^6 + \ldots \tag{4.9}$$

where a is arbitrary. Hence, retrieving ζ from the relation

$$\zeta \frac{dg}{d\zeta} = h - 2g, \tag{4.10}$$

$$f \sim -\frac{1}{6}\zeta^2 + A\zeta^3 - \frac{45}{2}A^2\zeta^4 + \ldots + \frac{aA^6}{5}\zeta^8 + \ldots, \tag{4.11}$$

where A is also arbitrary. If we now compare (4.11) with the general solution of (4.6a) for small ζ, namely

$$f \sim -\frac{1}{6}\zeta^2 + A\zeta^3 - \frac{45}{2}A^2\zeta^4 + \ldots + B\zeta^8 + \ldots \tag{4.12}$$

we see that the different phase curves through P_1 correspond to different choices of the second parameter B. There is only one choice of B which allows this solution to extend to $\zeta = +\infty$, and it is this choice which determines the powers of x and logx higher than x^8 in $d_o(x)$.

As $\zeta \to \infty$ along the phase curve P_1O, $h \sim 11/7$ so that $f \sim O(\zeta^{11/7})$ and the vapour concentration at $t=1/6$ is proportional to $-x^{11/7}$. We note that this is smaller than (4.3b) as $x \to -\infty$.

In $g > 0$, $h > 0$, there are an infinite number of phase curves with slope 11/7 emanating from O, where $\zeta \to -\infty$. All these trajectories give solutions with

$c \propto -x^{11/7}$ at $t = 1/6$. Consider first a trajectory such as OR in Fig. 3. ζ cannot vanish on such a trajectory, so that as R is approached $\zeta \to \zeta_o < 0$ and, from (4.10), $f \sim (\zeta_o - \zeta)^{\frac{1}{2}}$. Again, on OS, $\zeta \to 0-$ as S is approached with $g \sim \alpha_s/\zeta^2$, $h \sim -\beta_s/\zeta$ where α_s, β_s are constants. The corresponding function $f(\zeta)$ may be continued by beginning at S' where $g \sim \alpha_s/s^2$, $h \sim -\beta_s/\zeta$ as $\zeta \to 0+$. Finally, at S", $\zeta \to \zeta_1 > 0$ and $f \sim (\zeta_1 - \zeta)^{\frac{1}{2}}$. Repeating this procedure with OU takes us along U'O and f never vanishes for any finite ζ. It is only on OT and its continuation T'P$_2$ that $f \sim O(\zeta^* - \zeta)$ as $\zeta \to \zeta^* > 0$, and this happens as P_2 is approached. The different possibilities for f are shown in Fig. 4.

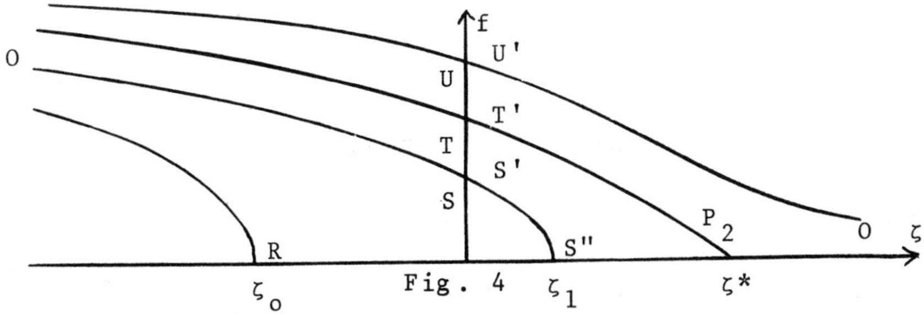

Fig. 4

We thus see that the condition (4.2) for a weak solution renders f unique and, near the moving boundary

$$c \propto t^{11/7} \,(\zeta^* - x/t^{7/3}) \qquad (4.13)$$

to lowest order.

There are other similarity solutions of the form $c = (t-1/6)^{2\beta-1} f\{x/(t-1/6)^{\beta}\}$, $\beta>0$. They correspond to $c_o - x^2$ having different series expansions from (4.4a) and suffer from the same deficiency as (4.5) in that they can only satisfy a singly infinite number of initial conditions in terms of a parameter analogous to A. However, they are all such that as $t \to 1/6 - 0$, $c \sim x^{2-1/\beta}$ and this suggests that it is only when c_o in (4.3b) grows at least as fast as x^2 as $x \to -\infty$ that c becomes singular in finite time as in (4.3a).

5. HELE-SHAW FLOW WITH SUCTION

We give a brief description of an intriguing singularity which can occur in Hele-Shaw flows with a moving boundary surrounding a point sink [14] and in analogous porous medium flows [12]. There may also be electrochemical applications [13].

A model for the pressure u in a Hele-Shaw cell from which liquid is being extracted at the origin O is

$$\Delta u = 0, \tag{5.1a}$$

$$u = 0 = u_t - L|\nabla u|^2 \text{ on } S(x,y,t) = 0 \tag{5.1b}$$

with $u \sim K \log r$ as $r^2 = x^2 + y^2 \to 0$. (5.2)

The moving boundary is $S = 0$ and is such that it surrounds O at $t = 0$ and $u(x,y,o)$ is compatible with this initial shape. L and K are positive constants so that (5.1,2) can be thought of as a Stefan problem for a supercooled liquid at temperature u with zero specific heat. When $S(x,y,o) = x^2 + y^2 - 1$, the solution is

$$S = x^2 + y^2 - (1-2LKt), \quad t < \frac{1}{2LK}, \tag{5.3}$$

but in [12] and [14] there are explicit solutions in which the boundary develops a cuspidal singularity in a finite time.

Problems of this type are apparently susceptible to the transformation [16]

$$\tilde{u} = \int_{\omega(x,y)}^{t} u(x,y,\tau)d\tau \tag{5.4}$$

where $S(x,y,\omega) \equiv 0$. This yields

$$\Delta\tilde{u} = L^{-1} \text{ with } \tilde{u} = \frac{\partial\tilde{u}}{\partial n} = 0 \text{ on } S = 0 \tag{5.5}$$

and $\tilde{u} \sim K(t-\omega_o)\log r$ as $r \to 0$ (5.6)

where ω_o, if it exists, is the time taken for the moving boundary to shrink to the origin. We appear to have removed the dependence of the solution on its past history and indeed, when $K<0$ and fluid is being injected at 0, (5-4)-(5-6) can be used as the departure point for an existence and uniqueness proof using linear complementarity ideas [5]. However, when say $S(x,y,o) \equiv x+1$, it is easy to construct solutions of the improper Cauchy problem (5.5) which give $\omega_o < 0$. This is because the steps leading to (5.5) break down if u or ω become singular.

That this can easily happen is shown in [12] and [14] where explicit solutions are found when $S(x,y,o) = 0$ is a circle whose centre is not at 0 or a straight line. Writing $u = K\log|\zeta|$ where $z=f(\zeta,t)$ maps $S=0$ into $|\zeta| = 1$, the simplest example is when $f = a_1(t)\zeta + a_2(t)\zeta^2$. Then the initial shape is a limaçon, although this also represents the later stages of development when $S(x,y,o) = (x-\varepsilon)^2 + y^2 - 1$ and $\varepsilon << 1$. Substituting f into (5.1b) gives

$$|a_1|^2 + 2|a_2|^2 = -2LKt, \quad a_1^2\,\bar{a}_2 = 1 . \tag{5.7}$$

Thus $2|a_2|^3 + 2LKt\,|a_2| + 1 = 0$ (5.8)
and $|a_2|$ is plotted in Fig. 5a, with the corresponding moving boundary positions in Fig. 5b. No solution exists for

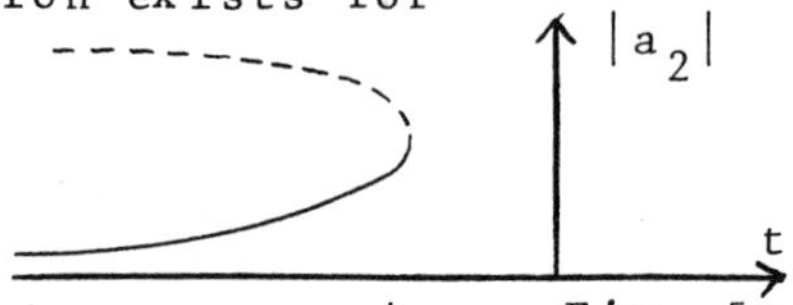

t_c Fig. 5a

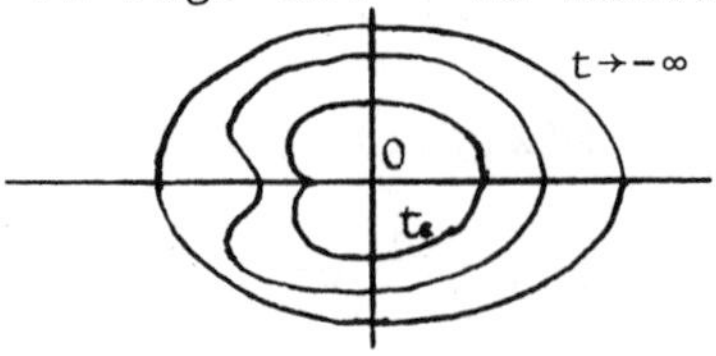

Fig. 5b

a_1 or a_2, real or complex for $t > t_c = -3/2^{4/3}LK$.

Unlike the problem in the previous section, the appearance of the boundary singularity is just one manifestation of the global breakdown of the solution. It is unlikely that incorporating the effects of surface tension, for example, will alleviate the situation. In view of the comment made after (5.2) the singularity is reminiscent of others which can occur in superheated or supercooled Stefan problems. Even in one space dimensions, it is easy to see that the boundary can move with infinite speed, as for example in the supercooled Neumann problem [3]. Sherman [17] has also exhibited such behaviour on finite domains. The possibility of such rapid boundary motion may have prompted Polubarinova-Kochina [12] to suggest that fluid inertia effects should be incorporated into such problems.

REFERENCES

1. A.A. Andronov, A.A. Vitt and S.E. Khaikin. Theory of Oscillations, Oxford, 1966.

2. D.R. Atthey. J.Inst.Maths Applics. 13 (1974), 353-366.

3. H.S. Carslaw and J. C. Jaeger. Conduction of Heat in Solids, Oxford University Press, London, 1959.

4. B. Chalmers. Principles of Solidification, Wiley, New York, 1964.

5. C.M.Elliott in Numerical Solution of Differential Equations, Vol. 4, ed. J. Albrecht, L. Collatz and G.Hammerling, Birkhauser-Verlag, Basel-Stuttgart, in press.

6. A.F. Filippov, Amer. Math. Soc. Transl. 42 (1960), 199-231.

7. J.L. Gravelleau and P. Jamet. SIAM J. Appl. Math. 20 (1971) 199-223.

8. E.A. Guggenheim. Thermodynamics, North-Holland, Amsterdam, 1967.

9. B. Knerr. Ph.D. Thesis, Northwestern U, 1976.

10. O.A. Oleinik. Sov. Math. Dokl. 1 (1960), 1350-1354.

11. O.A. Oleinik, A.S. Kalashnikov and Chzou Yui-Lin, Izv. Akad. Nauk. SSSR Ser. Mat. 22 (1958), 667-704.

12. P.Y. Polubarinova-Kochina. Theory of Ground Water Movement, Princeton, 1962.

13. H. Rasmussen and S. Christiansen. J. Inst. Maths Applics. 18 (1976), 149-153.

14. S. Richardson. J. Fluid Mech. 56 (1972), 609-618.

15. S. Rogers. Ph.D. Thesis, Oxford U, 1977.

16. A. Schatz. J. Maths. Anal. App. 28 (1969), 569-580.

17. B. Sherman. Quart. Appl. Math. 28 (1971), 377-382.

18. A.B. Tayler in Moving Boundary Problems in Heat Flow and Diffusion, ed. J.R. Ockendon and W.R. Hodgkins, Oxford 1974.

19. K. Taubert. Numer. Math. 26 (1976), 379-395.

J.R. Ockendon,
Oxford University Computing Laboratory,
19 Parks Road,
Oxford OX1 3PL, England.

The author would like to thank Drs. A.B. Crowley, C.M. Elliott and F.T. Smith for helpful discussions during the preparation of this paper.

FINITE ELEMENT METHODS FOR CERTAIN FREE BOUNDARY-VALUE PROBLEMS IN MECHANICS

J. T. Oden and N. Kikuchi

In this paper we consider a finite element formulation of several classes of free boundary-value problems in mechanics. In particular, we consider the contact problem in elasticity involving the contact of one elastic body on another. We also examine some one-phase Stephan problems encountered in thermafrost applications. In all of these problems, we develop formulations based on variational inequalities. We derive criteria for the selection of shape functions and also establish a priori error bounds. The results of numerical experiments are also presented.

1. INTRODUCTION

Our objective in this paper is to describe several finite element methods for the numerical solution of certain classes of free boundary problems encountered in continuum mechanics and in heat conduction. To keep the scope of this paper within reasonable bounds, we limit ourselves to methods based on formulations of various problems as variational inequalities, although we have also developed free-boundary methods which are not based on such formulations (e.g. WELLFORD and ODEN [15]). We note that a lengthy survey of methods for handling free- or moving-boundary problems has recently been compiled by CRYER [3] and that several accounts of applications of variational inequalities to such problems are available (see, e.g. DUVAUT and LIONS [5], GLOWINSKI [7], GLOWINSKI, LIONS and TREMOLIERES [8], and the references therein).

The problems we will consider fall into several broad categories. The standard variational inequality formulation is well known. Let K denote a non-empty, closed, convex subset of a real Hilbert space H, a: $H \times H \to \mathbb{R}$ a continuous, coercive, bilinear form on K --i.e., there exists positive constants γ and

ISBN 0-12-757350-X

M such that $\forall\, u,v \in K$

$$a(u,u) \geq \gamma \|u\|_{H}^{2}\ , \qquad a(u,v) \leq M\|u\|_{H}\|v\|_{H} \tag{1.1}$$

--and f a continuous linear functional of H. The standard variational inequality involves the problem of finding $u \in K$ such that

$$a(u,v-u) \geq f(v-u) \qquad \forall\, v \in K\ . \tag{1.2}$$

In all of the applications we consider, H will be a Sobolev space $H^m(\Omega)$, $m \geq 0$, with inner product and norm

$$(u,v)_m = \int_\Omega \Sigma_{|\alpha|\leq m} D^{\alpha} u\, D^{\alpha} v\, dx\ ; \qquad \|u\|_m^2 = (u,u)_m \tag{1.3}$$

or the subspace $H_0^m(\Omega)$ of $H^m(\Omega)$ consisting of functions whose derivatives of order less than or equal to $m-1$ vanish on $\partial\Omega$. Here Ω is an open bounded domain in $\mathbb{R}^n$ with boundary $\partial\Omega$ and $dx = dx_1 dx_2 \ldots dx_n$. For details on notations and on additional properties of these spaces, see ADAMS [1].

The applications we will consider involve variational inequalities more general than that implied by (1.2). In particular, we consider:

(i) <u>The One-Phase Stephan Problem</u>: Suppose that for fixed $x \in \Omega$ and every t, $0 \leq t \leq T$, $u(x,t) \in L^2(0,T)$ and for each t, $u(\underset{\sim}{x},t) \in H^1(\Omega)$. We then write $u \in L^2(0,T;H^1(\Omega))$. We consider here the problem of finding $u \in L^2\big(0,T;H^1(\Omega)) \cap K(t)\big)$ such that

$$\left.\begin{aligned} &\left(\frac{\partial u}{\partial t}, v-u\right) + a(u,v-u) \geq f(v-u) \qquad \forall\, v \in K(t) \\ &K(t) = \{v \in H^1(\Omega) : v|_{\partial\Omega} = g(t)\ ,\quad v \geq 0 \text{ a.e. in } \Omega\} \end{aligned}\right\} \tag{1.4}$$

Here $(\cdot,\cdot) = (\cdot,\cdot)_0$ is the L^2-inner product and

$$a(u,v) = \int_\Omega \nabla u \cdot \nabla v\, dx + \int_{\Gamma_F} \alpha u v\, ds\ , \qquad f(v) = \int_\Omega f v\, dx\ . \tag{1.5}$$

(ii) Contact Problems for Elastic Bodies. We will consider a model problem (the deflection of two parallel membranes) which falls into the general class of two-body contact problems: Find a pair of displacement vectors $\underset{\sim}{W} = (w^{(1)}, w^{(2)})$ such that

$$\underset{\sim}{W} \in K = \{\underset{\sim}{V} = (v^{(1)}, v^{(2)}) \in (H^1(\Omega_1))^3 \times (H^1(\Omega_2))^3 : \quad v^{(\alpha)}\big|_{\Gamma_D^{(\alpha)}} = 0 \ , \quad \alpha = 1,2 \ , \quad v_n^{(1)} + S \geq v_n^{(2)} \text{ on } \Gamma_c\} \tag{1.6}$$

such that

$$a(\underset{\sim}{W}, \underset{\sim}{V} - \underset{\sim}{W}) \geq f(\underset{\sim}{V} - \underset{\sim}{W}) \qquad \forall\, \underset{\sim}{V} \in K \tag{1.7}$$

where

$$\left.\begin{aligned} a(\underset{\sim}{W}, \underset{\sim}{V}) &= \Sigma_{\alpha=1}^{2}\, a_\alpha(\underset{\sim}{w}^{(\alpha)}, \underset{\sim}{v}^{(\alpha)}) \ ; \qquad f(\underset{\sim}{V}) = \Sigma_{\alpha=1}^{2}\, f_\alpha(\underset{\sim}{v}^{(\alpha)}) \\ a_\alpha(\underset{\sim}{w}^{(\alpha)}, \underset{\sim}{v}^{(\alpha)}) &= \int_{\Omega_\alpha} E_{(\alpha)}^{ijk\ell}\, \frac{\partial w_i^{(\alpha)}}{\partial x_j}\, \frac{\partial v_k^{(\alpha)}}{\partial x_\ell}\, dx \ , \quad \alpha = 1,2 \\ f_\alpha(\underset{\sim}{v}^{(\alpha)}) &= \int_{\Omega_\alpha} f_i^{(\alpha)} v_i^{(\alpha)}\, dx \ , \quad \alpha = 1,2 \ . \end{aligned}\right\} \tag{1.8}$$

Here Γ_c is the contact surface, $\Gamma_D^{(\alpha)} = \partial\Omega_\alpha - \Gamma_c$, $E^{ijk\ell}$ is Hooke's tensor of elastic constants, S is a prescribed boundary gap, $v_n^{(\alpha)}$ the component of $\underset{\sim}{v}^{(\alpha)}$ normal to Γ_c , and the summation convention is employed.

We remark that the rigid punch problem and Signorini's problem are both special cases of (1.8).

The fact that the variational formulations described by (1.4) -(1.8) are equivalent to the corresponding classical formulations of these problems (in an appropriate distributional sense) is easily verified using the arguments of, for example, DUVAUT and LIONS [5].

In the section following this Introduction, we discuss several general properties of finite element approximations of variational inequalities. We prove an existence theorem which leads to some results on the strong convergence of Galerkin approximations of a model two-body contact problem encountered in the study of deflections of parallel elastic membranes subjected to transverse pressures. Finally, in Section 5, we describe the results of several numerical experiments. We note that the results reported in this brief paper represent only portions of more extensive studies of problems of this type which are to be presented later by the authors.

2. FINITE ELEMENT APPROXIMATIONS

Let Ω be the domain of functions appearing in any of the variational inequalities mentioned in the Introduction. We construct finite element approximations of the spaces $H^m(\Omega)$ by following well-known techniques: We introduce a partition P_h of Ω into $E = E(h)$ subdomains (finite elements) $\{\Omega_e\}$ such that $\bar{\Omega} = \bigcup_{e=1}^{E(h)} \bar{\Omega}_e$, $\Omega_e \cap \Omega_f = \emptyset$, $e \neq f$, $h_e = \text{dia}\,(\Omega_e)$, and we represent the restrictions of functions in $H^m(\Omega)$ on each Ω_e by polynomials of degree k , $k > m \geq 0$, in such a way that each restriction is uniquely determined by specifying its values or the values of various derivatives at nodal points identified in the closure of each element. By constructions of this type, it is possible to develop a finite-dimensional subspace $S_h(\Omega)$ of $H^m(\Omega)$ for each partition P_h which has very useful interpolation properties.

In particular, suppose that each element Ω_e of the partition is the image, under an affine (or "almost" affine) invertible map of a fixed "master element" $\hat{\Omega}$. Let ρ_e = sup {diameters of all spheres inscribed in Ω_e }. Suppose, further, that all refinements of the finite element mesh are regular; i.e., there exists a constant $\sigma_o > 0$ such that $h_e/\rho_e \leq \sigma_o$, $h_e \to 0$ for all e , $1 \leq e \leq E(h)$, $h = \max\,(h_e)$. Then it is possible to establish the existence of a constant $C > 0$, independent of h and u , such that for any $u \in H^r(\Omega)$, $r > 0$,

$$\|u - \Pi_h u\|_m \leq Ch^\mu \|u\|_r$$

$$\mu = \min(k+1-m, r-m) \quad \Big\} \qquad (2.1)$$

where $\Pi_h : H^r(\Omega) \to H^m(\Omega)$ is a projection of $H^r(\Omega)$ onto $S_h(\Omega) \subset H^m(\Omega)$ generally endowed with the property that it preserves polynomials of degree k (i.e., $\Pi_h p = p \quad \forall\, p \in P_k(\Omega)$). Interpolation estimates such as (2.1) were derived by CIARLET and RAVIART [2] (for $r = k+1$) for both affine and isoparametric elements under the stated assumptions. We will assume that estimates such as (2.1) hold throughout this paper.

In view of (2.1), we see that finite element discretizations can be used to construct a family of subspaces $\{S_h(\Omega)\}$ of $H^m(\Omega)$, depending on the mesh parameter h, $0 < h \leq 1$, such that

$$\cup_{0<h\leq 1} S_h(\Omega) \text{ is everywhere dense in } H^m(\Omega) \qquad (2.2)$$

provided $\mu = \min(k+1-m, r-m) > 0$. We use such subspaces as a basis for approximation of problems such as (1.2) by adopting the following strategy:

(i) Let $\{S_h\}$ be a family of finite-dimensional subspaces of the real Hilbert space H of (1.2) such that $\cup_h S_h$ is dense in H (recall (2.2)). Construct a family of non-empty, closed, convex subsets $\{K_h\}$, $K_h \subset S_h$, such that $\forall\, v \in K \subset H$, a sequence $\{v_h\}$, $v_h \in K_h$, can be found which converges strongly to v in K and the weak limit u of the sequence $\{u_h\}$, $u_h \in K_h$, belongs to K. Note that in general $K_h \not\subset K$.

(ii) Seek a function $u_h \in K_h$ such that

$$a(u_h, v_h - u_h) \geq f(v_h - u_h) \quad \forall\, v_h \in K_h . \qquad (2.3)$$

Sufficient conditions for the existence of solutions of the approximate problem (2.3) are given in the following theorems.

<u>Theorem 1</u>. Let the following conditions hold:

(i) The bilinear form $a(\cdot,\cdot)$ is continuous and coercive on K_h in the sense that there exists a $v_0 \in K_h$ such that for every h , $|a(v_h, v_h - v_0)|/\|v_h\|_H \to +\infty$ as $\|v_h\|_H \to \infty$

(ii) $v \to a(v,v)$ is weakly lower semicontinuous; i.e., $\liminf_{h\to 0} a(v_h, v_h) \geq a(v,v)$, where $v_h \in K_h$ and $\{v_h\}$ converges weakly to $v \in K$.

Then there exists at least one solution $u_h \in K_h$ of (2.4) for each h . Moreover, there exists a subsequence $\{u_h'\}$ of the sequence of approximate solutions which converges weakly to the solution u of (1.2).

Further, suppose that in addition to (i) and (ii), there holds

(iii) Constants $C_1, C_2 > 0$ exist such that $\forall\, \varepsilon > 0$,

$$a(v,v) \geq C_1\|v\|_H^2 - C_2\|v\|_G^{1+\varepsilon} \qquad \forall\, v \in K \tag{2.4}$$

where $\|\cdot\|_G$ is the norm on a Hilbert space G with the property that H is compactly embedded in G . Then a subsequence $\{u_h''\}$ of approximate solutions can be found which converges strongly to the weak limit $u \in K$ of the sequence $\{u_h'\}$ described above.

Proof: By a well-known lemma due to HARTMANN and STAMPACCHIA [9], it is known that the problem $(Au, v-u) \geq f(v-u)$ $\forall\, v \in K$ has a solution $u \in K$ whenver A is continuous and K is a closed, bounded, convex subset of a finite-dimensional space H . Since $a(\cdot,\cdot)$ and f are continuous on $K_h \times K_h$, the existence of solutions to (2.4) easily follows.

By virtue of (i), the sequence $\{u_h\}$ of solutions obtained as $h \to 0$ is bounded and, therefore, contains a weakly convergent subsequence $\{u_h'\}$ with a limit $u \in K$. By (ii)

$$a(u,u) \leq \liminf_{h\to 0} a(u_h', u_h') \leq \liminf_{h\to 0} (a(u_h', v_h) - f(v_h - u_h'))$$

$$= a(u,v) - f(v,u) \quad .$$

Hence u satisfies (1.2) and is, therefore, a solution.

It remains to be shown that if (iii) holds, there exists a subsequence $\{u_h''\}$ which converges strongly to u . In accordance with (2.4),

$$0 \geq \lim_{h\to 0}\{a(u_h'',u_h''-u) - f(u_h''-u)\}$$

$$\geq \lim_{h\to 0}\{a(u \quad u_h''-u) - f(u_h''-u) + C_1\|u_h''-u\|_H^2 - C_2\|u_h''-u\|_G^{1+\varepsilon}\}$$

$$\geq C_1 \lim_{h\to 0}\|u_h''-u\|_H^2 \quad .$$

Hence $u_h'' \to u$ strongly in H. Here we have used the fact that $\|u_h''-u\| \to 0$ owing to the assumption that H is compact in G. ●

We have thus established that, under the rather mild conditions of Theorem 2.1, convergent finite element approximations of the solution of (1.2) can be constructed. To obtain some idea of the quality of such approximations, it is useful to consider a priori error estimates. Toward this end, we first introduce a general theorem due to FALK [6].

Theorem 2. Let (1.1) hold, $u \in K$ be the solution of (1.2) and $u_h \in K_h$ the solution of (2.3). Let $A : H \to H'$ be defined by

$$a(u,v) = \langle A(u),v\rangle$$

where $\langle\cdot,\cdot\rangle$ denotes duality pairing on $H' \times H$. Then, for every $v_h \in K_H$ and $v \in K$ we have

$$a(u-u_h,u-u_h) \leq a(u-u_h,u-v_h) + \langle f-Au,u-v_h+u_h-v\rangle . \quad (2.5)$$

Moreover, if $Au-f \in G$, G is the pivotal space which compactly contains H,

$$\|u-u_h\|_H^2 \leq \frac{M^2}{\gamma^2}\|u-v_h\|_H^2 + \frac{2}{\gamma}\|f-Au\|_G\left[\|u-v_h\|_G + \|u_h-v\|_G\right] \quad (2.6)$$

Proof: This theorem was proved by FALK [6]. We shall merely note that the proof follows directly from elementary operations on the inequality which follows from (1.2). ●

Remark 1. If $K_h \subset K$, we may take $v = u_h$, and (2.6) reduces to

$$\|u - u_h\|_H^2 \le \frac{M^2}{\gamma^2}\|u - v_h\|_H^2 + \frac{2}{\gamma}\|f - Au\|_G \|u - v_h\|_G \; . \qquad (2.7)$$

●

Remark 2. If $K = H$, we have further

$$\|u - u_h\|_H \le \frac{M}{\gamma}\|u - v_h\|_H \; , \qquad (2.8)$$

which is the standard estimate for linear elliptic equations (see, e.g., ODEN and REDDY [14]). ●

Remark 3. Suppose that $f - Au \in W'$ where W' is the dual of a Banach space W such that $G \subset W$. Then (2.5) holds with $\|f - Au\|_G$ and $\|u_h - v\|_G$ replaced by $\|f - Au\|_{W'}$, and $\|u_h - v\|_W$, respectively. This is a useful observation in cases in which $f - Au$ belongs to a non-reflexive space: e.g., $G = L^2(\Omega)$, $W = L^1(\Omega)$, $W' = L^\infty(\Omega)$, and $H = H^1(\Omega)$. In this case, we have

$$\|u - u_h\|_H^2 \le \frac{M^2}{\gamma^2}\|u - v_h\|_H^2 + \frac{2}{\gamma}\|f - Au\|_{W'}\left(\|u - v_h\|_W + \|u_h - v\|_W\right) \; . \qquad (2.9)$$

●

Remark 4. Consider the convex sets

$$K_1 = \{v \in H_0^1(\Omega) : |\mathrm{grad}\, v| \le 1 \text{ a.e. in } \Omega\}$$

$$K_2 = \{v \in H_0^1(\Omega) : v \ge \phi \text{ a.e. in } \Omega\}$$

and suppose $a(\cdot,\cdot)$ is given by (1.5) with $f \in C^\infty(\Omega)$. In each case, the maximum possible global rate of convergence in $H^1(\Omega)$ is $O(h)$; i.e., the optimal rate possible is achieved using piecewise linear elements (see FALK [6]). Hence, there seems to be little justification for using higher-order finite elements in such cases. ●

The question arises as to how the terms on the right side of inequality (2.5) can be evaluated for specific types of finite elements. We shall take up this question in specific applications to Stefan and contact problems in subsequent sections.

3. A STEFAN PROBLEM

Returning to (1.4), we now consider finite element approximations of a one-phase Stefan problem encountered in the study of thawing of a frozen media. In particular, we consider the problem of finding the potential $u(\underset{\sim}{x},t) \in L^2(0,T;H^1(\Omega))$, $\Omega \in \mathbb{R}^2$, satisfying

$$\left.\begin{aligned} & u\left(\frac{\partial u}{\partial t} - \Delta u - L\right) = 0 \ , \quad u \leq 0 \ , \quad \frac{\partial u}{\partial t} - \Delta u - L \leq 0 \quad \text{in} \quad \Omega \\ & u\Big|_{\Gamma_D} = \hat{g}(t) = \int_0^t \theta_o(\tau)\, d\tau \ , \quad \frac{\partial u}{\partial n}\Big|_{\Gamma_{\hat{F}}} = 0 \ , \\ & \frac{\partial u}{\partial n} + \alpha u\Big|_{\Gamma_F} = 0 \end{aligned}\right\} \tag{3.1}$$

where L is the latent heat, $\theta_o(t)$ the given boundary temperature on Γ_o , $\theta_o(t) < 0$, α a constant, and the boundary $\partial\Omega = \bar{\Gamma}_D \cap \bar{\Gamma}_{\hat{F}} \cap \bar{\Gamma}_F$ is composed of three segments on which the indicated conditions are prescribed. The temperature θ is given by

$$u(\underset{\sim}{x},t) = \int_{S(\underset{\sim}{x})}^t \theta(\underset{\sim}{x},\tau)\, d\tau \ . \tag{3.2}$$

The problem (3.1) is equivalent to (1.4), where $S(\underset{\sim}{x})$ is the time of arrival at the frozen front at the point $x \in \Omega$. The above formulation is due to DUVAUT [4].

A finite element approximation of (1.4) or (3.1) leads to the problem of finding $u_h(\underset{\sim}{x},t)$ such that

$$\left.\begin{aligned} & (\partial u_h, v_h - u_h) + a(u_h, v_h - u_h) \geq f(v_h - u_h) \quad \forall\, v_h \in K_h(t) \\ & K_h(t) = \{v_h \in S_h : v_h\big|_{\Sigma_h^D} = \hat{g}\big|_{\Sigma_h^D} \ , \quad v_h\big|_{\Sigma_h} \leq 0\ \} \ , \end{aligned}\right\} \tag{3.3}$$

where Σ_h^D and Σ_h denote the sets of all nodal points on Γ_o

and in Ω, and ∂ denotes the implicit finite difference operator for temporal approximations, i.e.

$$\partial u_h^n = \frac{1}{\Delta t}(u_h^{n+1} - u_h^n) \tag{3.4}$$

with Δt the time increment and $u_h^n = h_h(x,(n\Delta t))$.

We now consider error estimates for the above problem. To do this, we need an extension of the relationship (2.6) for the time dependent problem. We first establish a basic lemma.

Lemma 1. Let u and u_h be the solutions of the primal problem (3.1) and its approximation (3.3), respectively. Then, $\forall\, v \in K$,

$$\begin{aligned}(\partial e^n, e^{n+1}) + a(e^{n+1}, e^{n+1}) &\leq (\partial e^n, u^{n+1} - v_h) + \\ &+ a(e^{n+1}, u^{n+1} - v_h) + \langle \frac{\partial u^{n+1}}{\partial t} + Au^{n+1} - f^{n+1}, v - u_h^{n+1} + v_h - u^{n+1}\rangle \\ &+ (\partial u^n - \frac{\partial u^{n+1}}{\partial t}, v_h - u_h^{n+1}) \quad \forall\, v_h \in K_h((n+1)\Delta t),\end{aligned} \tag{3.4}$$

where $e^n = (u - u_h)^n$, $e^n = e(n\Delta t)$, and $\langle Au^{n+1}, v\rangle = a(u^{n+1}, v)$. ●

A proof of this lemma is to be published in KIKUCHI and ICHIKAWA [12]. Using Lemma 1, the following error estimate can be obtained, c.f. JOHNSON [10].

Theorem 3. Under the assumptions that $u \in L(0,T;H^{(5/2)-\varepsilon}(\Omega))$, $\frac{\partial u}{\partial t} \in L^2(0,T;H^2(\Omega))$, $\frac{\partial u}{\partial t} + Au - f \in L(0,T;L^\infty(\Omega))$, $\varepsilon > 0$, and that the speed of propagation of frozen front is of order t^α (α may be negative), the following estimate holds:

$$\operatorname{Max}_n \|e^n\|^2 + \hat{\gamma}\|e^{n+1}\|^2 \Delta t \leq C(u, \frac{\partial u}{\partial t}) h\gamma + C\Delta t^\mu . \tag{3.5}$$

Here $\mu = \min(2, 1 + 2(1+\alpha))$ and

$$\left.\begin{aligned}\gamma &= 2 \quad \text{for linear finite elements} \\ \gamma &= 3 - \varepsilon \quad \text{for quadratic finite elements}\end{aligned}\right\} \tag{3.6}$$

Proof: We only sketch the essential details of the proof; for complete details, see KIKUCHI and ICHIKAWA [12]. From Lemma 1, it suffices to estimate the right-hand side of (3.4).

(i) For the term $(\partial e^n, u^{n+1} - v_h)$, we have

$$\Sigma_{n=0}^{N-1} (\partial e^n, u^{n+1} - v_h)\,\Delta t \leq \varepsilon \Sigma_{n=0}^{N-1} \|e^{n+1}\|_0^2\,\Delta t + C_1 h^{(2,4)} \left\|\frac{\partial u}{\partial t}\right\|^2_{L^2(0,T;H^2(\Omega))} + \varepsilon\|e^N\|_0^2 + \varepsilon\|e^0\|_0^2 + C_2 h^{(2,4)} \|u\|^2_{L^2(0,T;H^2(\Omega))} .$$

(ii) For $a(e^{n+1}, u^{n+1} - v_h)$, we have

$$\Sigma_{n=0}^{N-1} a(e^{n+1}, u^{n+1} - v_h)\,\Delta t \leq \varepsilon \Sigma_{n=0}^{N-1} \|e^{n+1}\|_1^2\,\Delta t + C_3 h^{(2,3-2\varepsilon)} \|u\|^2_{L^\infty(0,T;H^{\hat\beta}(\Omega))} ,$$

where $\hat\beta = 2$ for linear elements, $\hat\beta = \frac{5}{2} - \varepsilon$ for quadratic elements.

(iii) For $(\partial u^n - \frac{\partial u^{n+1}}{\partial t}, v_h - u_h^{n+1})$, we have, under the stated condition on the speed of propagation,

$$\Sigma_{n=0}^{N-1} (\partial u^n - \frac{\partial u^{n+1}}{\partial t}, v_h - u_h^{n+1})\,\Delta t \leq \left\|\partial u - \frac{\partial u}{\partial t}\right\|_{L^\infty(0,T;L^2(\Omega))} \Delta t^2 + C_4 h^{(2,4)} \|u\|^2_{L^\infty(0,T;H^2(\Omega))} + 2\varepsilon \Sigma_{n=0}^{N-1} \|e^{n+1}\|^2\,\Delta t + C_5 \left\|\frac{\partial u}{\partial t}\right\|^2_{L^2(0,T;H^1(\Omega))} \Delta t^2 + C_6 \Delta t^{1+2(\alpha+1)}$$

(iv) For the last part,

$$\Sigma_{n=0}^{N-1} \left\langle \frac{\partial u^{n+1}}{\partial t} + Au^{n+1} - f^{n+1}, v - u_h^{n+1} + v_h - u^{n+1} \right\rangle \Delta t \leq C_7 \Delta t\, h^{(2,3)}.$$

Here $h^{(\alpha,\beta)}$ signifies that the exponent of h is α for linear finite elements and β for quadratic elements. Combining all of the above estimates gives (3.5). ●

4. CONTACT PROBLEMS

We shall confine ourselves to a simple model problem; we refer the reader interested in full details of general problems of this type to the work of KIKUCHI [11].

A conceptually simple model of the two-body contact problem is embodied in the description of the deflection of two parallel membranes loaded laterally by pressures $f_\alpha = f_\alpha(x_1,x_2)$, $\alpha = 1,2$, as shown in Fig. 1. The variational inequality characterizing this problem is as follows: Find $\underset{\sim}{W} = (w^{(1)},w^{(2)}) \in K$ such that

$$\left.\begin{aligned}
&\int_{\Omega_1} T_1 \nabla w^{(1)} \cdot \nabla (v^{(1)} - w^{(1)})\, dx + \int_{\Omega_2} T_2 \nabla w^{(2)} \cdot \nabla (v^{(2)} - w^{(2)})\, dx \\
&\qquad \geq \int_{\Omega_1} f_1 (v^{(1)} - w^{(1)})\, dx + \int_{\Omega_2} f_2 (v^{(2)} - w^{(2)})\, dx \\
&\qquad \forall\ (v^{(1)},v^{(2)}) \in K \\
&K = \{(v^{(1)},v^{(2)}) \in H_0^1(\Omega_1) \times H_0^1(\Omega_2) :\ v^{(1)} + \phi \geq v^{(2)} \quad \text{a.e. in } \Omega\}
\end{aligned}\right\} \tag{4.1}$$

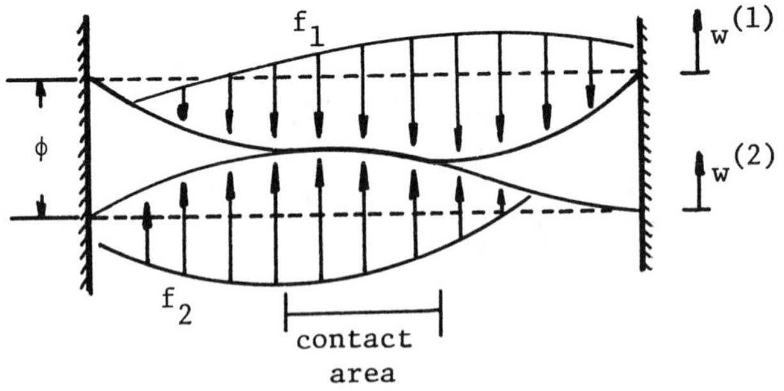

Figure 1. Contact problem for two parallel membranes subjected to transverse pressures.

Here T_1 and T_2 are tensions in membranes 1 and 2, respectively, and are known constants, $\phi = \phi(x_1, x_2)$ is the initial gap between the unloaded membranes. The terms on the right side of (4.1) define a continuous, coercive bilinear form $a(\underset{\sim}{W}, \underset{\sim}{V}) = a_1(w^{(1)}, v^{(1)}) + a_2(w^{(2)}, v^{(2)})$ on $H_o^1(\Omega_1) \times H_o^1(\Omega_2)$, where $a_\alpha(\cdot,\cdot)$ is given by (1.5) with integration understood over Ω_α, $\alpha = 1,2$. This problem, therefore, has a unique solution $\underset{\sim}{W} \in K$.

Let us now consider a finite element approximation of (4.1) obtained by partitioning each Ω_α into triangles over which piecewise linear approximations are used for components $w^{(1)}$ and $w^{(2)}$. Let

$\Sigma_h^{(\alpha)}$ = the set of all nodal points in the partition of Ω_α, $\alpha = 1,2$

$\Gamma_h^{(\alpha)}$ = the set of all nodal points on the boundaries $\partial\Omega_\alpha$, $\alpha = 1,2$.

Each domain Ω_α is assumed to be represented exactly by the union of the triangles making up these parallel finite element meshes. The approximation of (4.1) is characterized by the problem of finding $\underset{\sim}{W}_h = (w_h^{(1)}, w_h^{(2)}) \in K_h$ such that

$$a(\underset{\sim}{W}_h, \underset{\sim}{V}_h - \underset{\sim}{W}_h) \geq \Sigma_{\alpha=1}^{2} \int_{\Omega_\alpha} f_\alpha (w_h^{(\alpha)} - v_h^{(\alpha)})\, dx \qquad \forall\, \underset{\sim}{V}_h \in K_h ,$$

where

$$K_h = \Big\{ (v_h^{(1)}, v_h^{(2)}) \in S_h(\Omega_1) \times S_h(\Omega_2) : v_h^{(1)}(\sigma_h^{(\alpha)}) + \phi(\sigma_h^{(\alpha)}) \geq v_h^{(2)}(\sigma_h^{(\alpha)}) \quad \forall\, \sigma_h^{(\alpha)} \in \Sigma_h^{(\alpha)} ,\ \alpha = 1 \text{ and } \alpha = 2 \Big\} .$$

In other words, for the finite element approximation we apply the constraint at each node; e.g., if $\underset{\sim}{\zeta}$ is the coordinate of a node in the mesh triangulating Ω_1, we enforce the condition $v_h^{(1)}(\underset{\sim}{\zeta}) + \phi(\underset{\sim}{\zeta}) \geq v_h^{(2)}(\underset{\sim}{x})$, where $\underset{\sim}{x}$ is the point in Ω_2 corresponding to (below) $\underset{\sim}{\zeta}$. The scheme is illustrated in Fig. 2a.

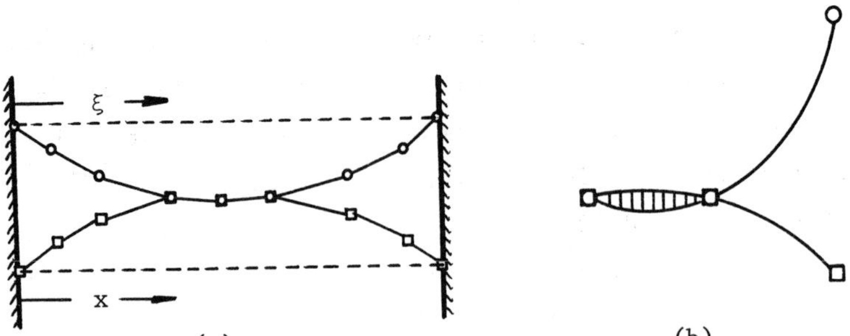

Figure 2. Imposition of constraints in finite element model.

Clearly, $K_h \subset K$, so that by virtue of Theorem 2 and Remark 1 following it, we have the error estimate

$$\|W - W_h\|_1^2 = \|w^{(1)} - w_h^{(1)}\|_{1,\Omega_1}^2 + \|w^{(2)} - w_h^{(2)}\|_{1,\Omega_2}^2$$

$$\leq \frac{M}{\gamma}\|W - W_h\|_1^2 + \frac{2}{\gamma}\|f - A(W)\|_0 \|W - V_h\|_0$$

$$\leq C_1 h^2 \|\underset{\sim}{W}\|_2^2 + C_2(\underset{\sim}{W}) h^2 \|\underset{\sim}{W}\|_2 \ .$$

Hence

$$\|\underset{\sim}{W} - \underset{\sim}{W}_h\|_1 \leq C(\underset{\sim}{W}) h \quad \text{as} \quad h \to 0 \ . \tag{4.2}$$

Notice that for the quadratic elements (see Fig. 2b), K_h is not contained in K and we must then work with the more general estimate (2.9). To obtain an error estimate for this case, we must estimate terms such as $\|\underset{\sim}{W}_h - \underset{\sim}{v}\|_{\mathcal{W}}$ for an appropriate product space $\mathcal{W}$. In this case, it is possible to show (see KIKUCHI [11]) that under the assumption

$$w^{(\alpha)} \in W^{2,\infty}(\Omega_\alpha) \ ; \quad f_\alpha \in L^\infty(\Omega_\alpha) \ , \quad \phi = \text{constant} \ , \tag{4.3}$$

we have

$$\|f - Au\|_{\mathcal{W}} \|u_h - v\|_{\mathcal{W}'} \leq Ch^2 \ ,$$

since the situation described in Fig. 2b may happen. Then the final estimate, given by (2.12) is

$$\|\underset{\sim}{W}-\underset{\sim}{W}_h\|_1^2 \leq C_1(\underset{\sim}{W})h^4 + (C_2(\underset{\sim}{W})h^4 + C_3(\underset{\sim}{W})h^2) \quad , \quad \text{or}$$

$$\|\underset{\sim}{W}-\underset{\sim}{W}_h\|_1 = O(h) \quad \text{as} \quad h \to 0 \ . \tag{4.3}$$

Thus, we gain no improvement in the rate of convergence by using quadratic or higher-order finite elements. The maximum rate of convergence in $H^1(\Omega_1) \times H^1(\Omega_2)$ is obtained by piecewise linear approximations, and the error in this norm is $O(h)$.

5. NUMERICAL RESULTS

We now show several numerical examples of problems of the type described previously.

(i) Contact Problems: We describe a numerical experiment for one-dimensional domains. In the formulation (4.1), we take $T^{(1)} = T^{(2)} = 1$, $f^{(1)} = -8$, $f^{(2)} = 6$, $\phi = 1$, $\Omega = (0,1)$. Then, the exact solution is

$$\begin{cases} \left.\begin{aligned} w^{(1)} &= 4x^2 - (7a - 0.5)x \\ w^{(2)} &= -3x^2 + (7a - 0.5)x \end{aligned}\right\} & 0 \leq x \leq \frac{1}{\sqrt{7}} \\ w = w^{(1)} = w^{(2)} = \frac{1}{2}x^2 - \frac{1}{2}x + \frac{1}{2} & \frac{1}{\sqrt{7}} \leq x \leq \frac{1}{2} \end{cases} ,$$

where $a = \frac{1}{\sqrt{7}}$ (see Fig. 3). The numerical results confirm the theoretical estimate (4.2); i.e., the rate of convergence is $O(h)$ for linear finite element approximations.

(ii) One-phase Stefan problem: We describe numerical experiments designed to establish actual rates of convergence for a one-phase Stefan problem. As shown in Figure 4, the rate of convergence in H^1-norm is exactly $O(h)$, but the rate in L^2-norm is only $O(h^{1.7})$ for the linear finite element approximation. For the quadratic elements, we also solved a "stationary" problem and obtained an $O(h^{1.5})$ error in the H^1-norm, $O(h^2)$ in the L^2-norm, as indicated in Figure 5. Finally, an example of a two-dimensional one-phase Stefan problem is shown in Figure 6.

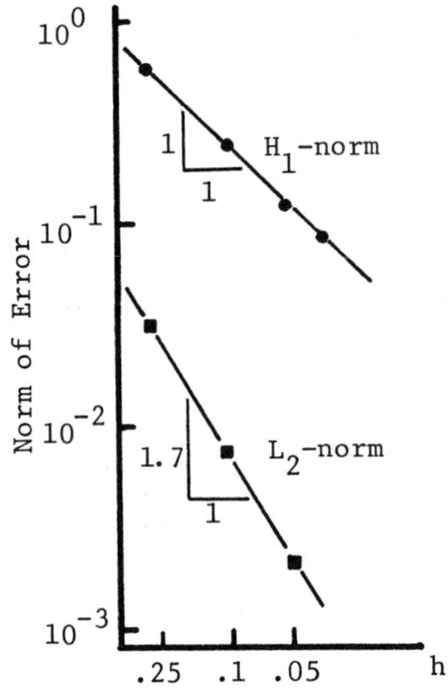

Figure 3.
Estimates for Two-String
Contact Problem

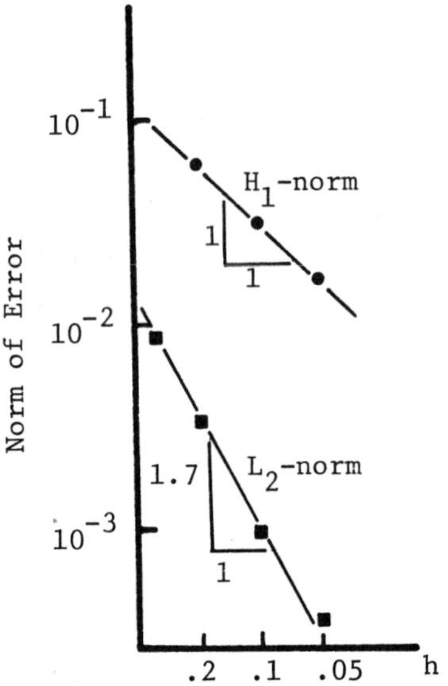

Figure 4.
Estimates for Stefan Problem

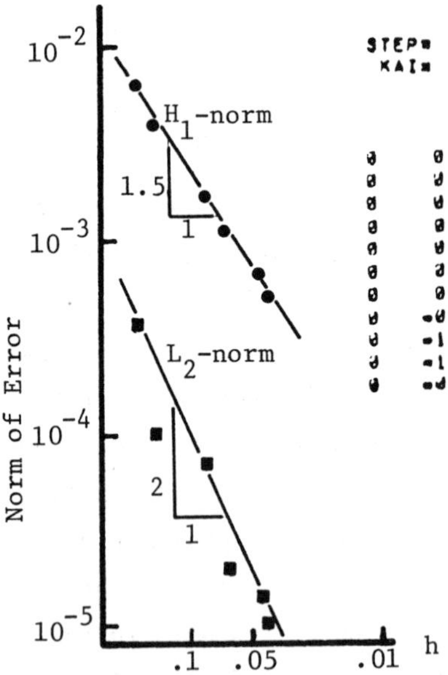

Figure 5.
Estimates for Stationary Stefan Problem

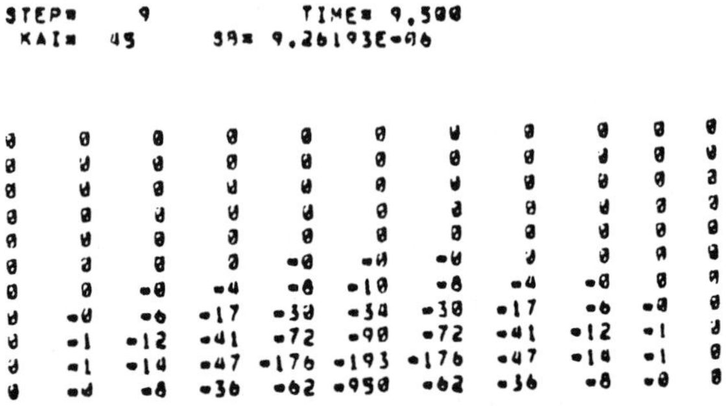

STEP= 9 TIME= 9.500
KAI= 45 SS= 9.26193E-06

0	0	0	0	0	0	0	0	0	0	0
0	0	0	0	0	0	0	0	0	0	0
0	0	0	0	0	0	0	0	0	0	0
0	0	0	0	0	0	0	0	0	0	0
0	0	0	0	0	0	0	0	0	0	0
0	0	0	0	-0	-0	-0	0	0	0	0
0	0	-0	-4	-8	-10	-8	-4	-0	0	0
0	-0	-6	-17	-30	-34	-30	-17	-6	-0	0
0	-1	-12	-41	-72	-90	-72	-41	-12	-1	0
0	-1	-14	-47	-176	-193	-176	-47	-14	-1	0
0	-0	-8	-36	-62	-950	-62	-36	-8	-0	0

Figure 6.
An Example of the Two-
Dimensional Stefan Problem

REFERENCES

1. R. A. Adams. Sobolev Spaces, Academic Press, 1975.

2. P. G. Ciarlet and R. A. Raviart. "General Lagrange and Hermite Interpolation in $\mathbb{R}^n$ with Application to the Finite Element Method," Arch. Rat. Mech. Anal., 46, (1972), 177-199.

3. C. W. Cryer. A Survey of Trial Free-Boundary Methods for the Numerical Solution of Free Boundary Problems, MRC Technical Summary Report #1963, University of Wisconsin, 1976.

4. G. Duvaut. "Résolution d'un Problème de Stefan," C.R.A.S., 276, (1973), 1461-1463.

5. G. Duvaut and J. L. Lions. Inequalities in Mechanics and Physics, Translated by C. W. John, Springer-Verlag, 1976.

6. R. S. Falk. "Error Estimates for the Application of a Class of Variational Inequalities," Math. Comp., 28, (1974), 963-971.

7. R. Glowinski. "Introduction to the Approximation of Elliptic Variational Inequalities," C.N.R.S., Universite Paris VI, 1976.

8. R. Glowinski, J. L. Lions and R. Trémolières. Analyse Numérique des Inéquations Variationnelles, Dunod, 1976.

9. P. Hartman and G. Stampacchia. "On Some Non-linear Elliptic Differential-Functional Equations," ACTA. Math., 115, (1966), 271-310.

10. C. Johnson. "A Convergence Estimate for an Approximation of a Parabolic Variational Inequality," SIAM J. Numer. Anal., 13, (1976), 599-606.

11. N. Kikuchi. "Contact Problems by Variational Inequalities," (to appear).

12. N. Kikuchi and Y. Ichikawa, "A Stefan Problem by Variational Inequalities," (to appear).

13. J. L. Lions. Quelque Méthodes Resolution des Problèmes aux Limites non Linéaires, Dunod, 1969.

14. J. T. Oden and J. N. Reddy. An Introduction to the Mathematical Theory of Finite Elements, John-Wiley & Sons, 1976.

15. L. C. Wellford, Jr. and J. T. Oden. "Discontinuous Finite-Element Approximations for Analysis of Shock Waves in Nonlinear Elastic Materials," J. Comp. Phy., 19, (1975), 179-210

J. T. Oden and N. Kikuchi
Texas Institute for Computational Mechanics
The University of Texas at Austin
Austin, Texas 78712

The work reported in this communication was supported by Grant AROD DAAG 29-77-G0087 from the U. S. Army Research Office-Durham.

MOVING BOUNDARY PROBLEMS

COMPARISON OF NUMERICAL METHODS FOR DIFFUSION PROBLEMS WITH MOVING BOUNDARIES

N. Shamsundar

Recent progress in methods for obtaining numerical solutions of Stefan type problems is reviewed, with attention focused on multidimensional problems. Comparisons are made between different ways of spatial and timewise discretization, choice of either enthalpy or temperature as the primary variable, and between solutions carried out in physical space or in transform space. Some unpublicised characteristics of the enthalpy based methods are illuminated. Guidelines are provided for selecting computational techniques for specific areas of application. Areas needing emphasis in future research work are indicated.

1. INTRODUCTION

For several years, diffusion problems in media containing a moving interface have received steady attention, and the emergence of new applications has recently given a marked impetus to their study. Most of the older investigations confined themselves to one-dimensional problems, and an arsenal of analytical and computational techniques have been established. Indeed, it is safe to say that one-dimensional problems are amenable to efficient and accurate numerical solution, with only moderate effort. Multidimensional Stefan problems are of more technological importance, but much harder to solve than one-dimensional problems Most of the powerful techniques usable for one-dimensional problems do not admit extension to higher dimensions. However, a few general solution methods for multidimensional problems do exist, some of them of recent creation. The intent of this paper is to give an overview of these and to make a comparative study of their features.

ISBN 0-12-757350-X

After giving the classical statement of the general problem, we define a few important physical parameters that have considerable impact on the selection of a technique for obtaining a solution. Then, the several methods of attack available are described, and their unique features are pointed out. This is followed by a comparative discussion of all these methods.

To facilitate the discussions, we employ the terminology that applies to problems where melting and solidification occur. Since many problems with moving boundaries are similar to these, application of the methods described here to other areas can be inferred by straightforward substitution of suitable terminology.

2. CLASSICAL STATEMENT OF THE PROBLEM

At some instant of time t, there are two regions V_s and V_ℓ which adjoin each other. Each of these regions contains one phase of a pure substance. For concreteness, we refer to the phases as solid and liquid, although it is conceivable that another solid phase or a gaseous phase may be present instead. The surface Σ which separates these two regions, namely, the interface or phase boundary, moves as time proceeds. It is assumed that the union V of V_s and V_ℓ does not vary with time. Most of the time, we also assume that there is no flow of matter across the bounding surface of V, although this assumption must be cast aside when one cannot neglect the difference in density between the two phases. Similarly, it is assumed that there is no convective motion within V. The motion of the interface, the changes in temperature, etc. are governed by the conditions existing on the surface S of V, the initial distribution of substance and temperature, and upon the properties of the substance. At some instant of time, the interface Σ can cross S, and either appear for the first time, or disappear. Correspondingly, V_s or V_ℓ can be nonexistent during certain periods.

The heat equation in the solid is (see Nomenclature)

$$\rho_s c_s \, \partial T/\partial t = \nabla \cdot (k_s \nabla T) \tag{2.1}$$

and in the liquid

$$\rho_\ell c_\ell \, \partial T/\partial t = \nabla \cdot (k_\ell \nabla T) \tag{2.2}$$

At the interface, the temperature has to equal the equilibrium phase change temperature, T_{sat}, and the temperature distributions in the two single-phase regions are coupled by the energy conservation condition

$$(k \, \partial T/\partial n')_s - (k \, \partial T/\partial n')_\ell = \rho_s \lambda \, v'_\Sigma \tag{2.3}$$

where n' is the distance along the local normal to Σ and v'_Σ is the velocity of Σ along n'. At the external boundary S, boundary conditions of the type

$$p \, \partial T/\partial n' + qT + r = 0 \tag{2.4}$$

apply, where p,q,r vary with position and time, sometimes depending also on the local temperature.

If it is not permissible to ignore motion of the liquid, whether caused by buoyancy or by difference in density between the two phases, $\partial T/\partial t$ in (2.2) should be replaced by the substantial derivative DT/Dt. In this case, (2.3) remains valid, but evaluating the term $-(k\partial T/\partial n)_\ell$ involves solving the thermal convection problem in the liquid, which occurs in the presence of a moving boundary. A way of skirting this difficulty, one used often, is to replace $-(k\partial T/\partial n)_\ell$ by $h\,(T_{sat}-T_{liq})$; h is a heat transfer coefficient and T_{liq} is the bulk mean temperature of the assumedly well-mixed liquid. In most of the work to date, the correlations used for calculating h were those developed for similar geometries involving a stationary boundaries.

The equations (2.1) - (2.3) are linear in the temperature T, but the location of the interface is an unknown function of time. In fact, the primary result of solving a Stefan problem is usually the interface location itself. However, in some of the newer applications the interface is of secondary interest, other quantities such as surface temperatures and fluxes being the prime candidates for study.

3. PARAMETERS GOVERNING PHASE CHANGE

There are several ways of nondimensionalizing the basic equations, but the one described below has several advantages. The latent heat will appear only in conbination with the sensible heat terms of equations (2.1) and (2.2). In many problems, the so-called quasistationary problems, sensible heat has little contribution to the energy balance. When such is the case, the solutions for very different combinations of properties are repsentable by a single graph or equation involving the proper nondimensional variables.

Let $\vec{r} = \vec{r}'/L$, $\theta = (T-T_{sat})/(T_0-T_{sat})$, τ = Ste Fo $= [c_s(T_0 - T_{sat})/\lambda]\,[(k_s/\rho_s c_s)t/L^2]$ be the nondimensional position, temperature and time variables. Let us also assume, merely to preserve clarity in the discussion, that <u>the properties of each phase are uniform and constant in time.</u> Then, the nondimensional equivalents of our equations (2.1) - (2.4) are

$$\text{Ste}\ \partial\theta/\partial\tau = \nabla^2\theta \tag{3.1}$$

$$\text{Ste}\ (\rho_\ell/\rho_s)\ (k_s/k_\ell)\ \partial\theta/\partial\tau = \nabla^2\theta \tag{3.2}$$

$$(\partial\theta/\partial n)_s - (\partial\theta/\partial n)_\ell = dn/d\tau \tag{3.3}$$

$$(T_0-T_{sat})\ /\ L\ p\ \partial\theta/\partial n + (T_0-T_{sat})\ q\theta + (qT_{sat} + r) = 0 \tag{3.4}$$

Here, T_o is some external reference temperature such that $(T_o - T_{sat})$ is a characteristic temperature difference for the moving boundary problem, and Ste is the Stefan number based on this temperature difference.

The Stefan number $Ste = c_s(T_o - T_{sat})/\lambda$ is the parameter that identifies the importance of sensible heat as compared to latent heat. Note that it multiplies the sensible heat term in each of the heat equations (3.1) and (3.2), and appears nowhere else. In fact, the very general boundary condition (3.4) has the same form as (2.4); neither it nor the interfacial condition (3.3) involves Ste. Now then, the advantage of our formulation is evident. If sensible heat effects are small, Ste is small; for a quasistationary problem, (3.1) and (3.2) reduce to the Laplace equation, and the problem becomes independent of Ste. The influence of latent heat on the problem is solely due to the appearance of λ in the definition of τ. For small values of Ste, the dependence of θ on $\vec{r}$ and τ is barely influenced by Ste.

We have now indicated that when the Stefan number is small, it affects the resulting solution very little. On the other hand, many numerical techniques for solving the set (3.1) - (3.4) are greatly affected as to their efficiency of computation by the value of Ste, as we shall see later. As a consequence, with such techniques a computation for Ste = 0.01 may take ten times the effort needed with Ste = 0.1, even though the results may differ by only a couple of percent. This will be illustrated in more detail later.

From the foregoing remarks, the need for investigating the values of Ste that arise in various classes of problems is evident. By surveying diverse application areas and by considering a large variety of substances, the following estimates of Ste may be obtained.

APPLICATION	RANGE OF Ste
Casting and processing pure metals, with a large temperature change	1 to 3
Casting and processing alloys, with a large temperature change	0.2 to 3
Thermal energy storage; freezing and melting of water by Nature; both with a small temperature range	0 to 0.2

Additional parameters that may appear in phase change problems are the ratios (ρ_ℓ/ρ_s), (k_ℓ/k_s), the Biot number Bi (hL/k), etc.

4. SOLUTION METHODS BASED ON TEMPERATURE

By far the largest number of existing solutions for multidimensional phase change problems involve a numerical solution of the equation set (2.1) - (2.4) with no further modifications. The solution domain is discretized by employing a fixed grid of lines, and the nodal temperatures are obtained by solving the finite difference equations corresponding to (2.1) - (2.4). So far, only the standard explicit difference scheme has been used in various forms. Special equations need to be written in the neighborhood of the curved interface to account for the interfacial energy balance. Since equation (2.3) contains the temperature gradient normal to the interface, it is difficult to develop a finite difference equivalent for it that retains the second order accuracy with respect to the spatial step-size that one has at a single-phase node, unless a special, semianalytical starting solution is at hand. By extending the Murray-Landis explicit difference technique [13] to two dimensions, Springer and Olson [20], Rathjen and Jiji [14], Bilenas and Jiji [2] and Tien and Wilkes [21] obtained solutions to several two-dimensional problems. Lazaridis [10] recast equation (2.3) by expressing $\partial T/\partial n$ in terms of derivatives $\partial T/\partial x$ and $\partial T/\partial y$ along the coordinate axes. He then obtained solutions to several two-dimensional phase change problems and a three-dimensional problem. In references

[2] and [21], liquid motion is also calculated. Although the problem considered in [21] involves only steady state, the method of false transients was employed to obtain the steady state solution - one solves the transient equation with arbitrary initial conditions and performs calculations until steady-state is reached.

The only significant drawback of the explicit difference technique is the necessity to place restrictions on the maximum time-step to avoid numerical instability. For example, the stability condition at an interior node for equation (3.1) is $\Delta\tau \leq \text{Ste}\ \Delta^2/S$ where $\Delta\tau$ and Δ are the step sizes in τ and $\vec{r}$; S=2, 4 and 8 for one, two and three dimensional problems, respectively. Even though two problems, one with Ste = 0.1 and the other with Ste = 0.01, may have nearly identical solutions, the stability criterion forces one to use a $\Delta\tau$ for Ste = 0.01 that is one tenth that for Ste =0.1. When the difference equations corresponding to boundary conditions are considered, one sees again that stability criteria of the same form need to be satisfied; S may be smaller than the values given above. A penalty is also paid when Δ is reduced to improve the accuracy of the results. If Δ is halved, for example, the error is reduced by one-fourth, but the effort is increased sixteen times. Despite the simplicity of the explicit method, the stability problem may preclude its use for all problems other than those in which Ste is high and the time-span of the transient is small (for example, casting of metals).

There has been no work, to the knowledge of the author, in which an implicit difference scheme together with a fixed grid had been used. Such a discretization would have the advantage of stability offset by the need to solve a complicated set of nonlinear algebraic equations. There is a need for research in this direction.

The elegant method of invariant imbedding developed by Meyer is described by him in another paper in this monograph. Meyer has described a two-dimensional algorithm based on invariant imbedding [11] which is similar to the ADI method. In this algorithm, advancing the solution one step in time over a rectangular grid of m x n nodes involves the solution of 2(m + n) initial value problems for second order ordinary differential equations. At present, the technique cannot accommodate interfaces that cross any line parallel to the coordinate axes more than once.

Another class of methods is based on retaining temperature as the dependent variable, but the spatial variables are transformed so as to "immobilize" the moving interface. In the transformed space, the solid and liquid each occupy a fixed region, but the governing differential equations contain the parameters of the interface location. As a result of the immobilization, finite difference equations are easily written near the interface, but the equations themselves are more complicated, and must be solved iteratively. Spaid et al [19] made such a transformation and applied a modified explicit difference scheme which is, apparently, unconditionally stable. Our own experiments with such a scheme showed that instability is unavoidable when convective boundary conditions are specified. A slightly different immobilization was used by Duda et al [5], who used a fully implicit difference scheme to solve an axisymmetric cylindrical phase change problem. In solving the nonlinear algebraic equations resulting form the finite diffenerce representation, the authors had to employ an iterative scheme involving two levels of iteration for each step. Notwithstanding this complexity, the implicit scheme was still superior to the explicit method with its stability problems. In a more recent paper, Saitoh [15] sets forth a novel immobilization scheme that employs polar coordinates regardless of the shape of the body, and makes efficient treatment possible not only of the interface, but of arbitrarily shaped stationary boundaries. If the interface is given by $r = F(\phi,t)$, the external boundary by $r = B(\phi)$, the new variables are

$\eta = (r-F)/(B-F)$ and Φ, so that $\eta = 0$ at the interface, and $\eta = 1$ at the boundary, regardless of ϕ; thus, an arbitrary region is mapped into a rectangle. Subsequent to immobilization, or "boundary fixing", as Saitoh calls it, the resulting equations were solved with an implicit difference scheme.

For treating problems with boundaries on which the temperature is specified, the "isotherm migration method" is attractive. This method consists of exchanging one of the spatial variables with temperature, making the former the dependent variable and the temperature one of the independent variables. Crank and Gupta [4] used the standard explicit difference method to solve two-dimensional phase change problems. If a uniform mesh is employed in the new variables, the corresponding mesh in the physical plane has the desirable property that it is finer in regions with large temperature gradients, and vice versa. This advantage is thwarted by the difficulties in applying the method to situations where boundary temperatures are unknown, and by the inapplicability of implicit difference schemes.

It was noted in Section 3 that, when the Stefan number is small, the problem reduces to one of solving Laplace equations coupled by the interfacial conditions $T = T_{sat}$ and equation (3.3). When such is the case, the powerful array of techniques for solving the Laplace equation come within reach; for example, conformal transformations may be applied, as shown by Siegel et al. [18]. When the boundary conditions are either adiabatic or isothermal, this method becomes semianalytical in nature. But, for more general boundary conditions, the transformation would have to be carried out numerically. These methods apply, a fortiori when the problem involves a free boundary but not a moving one, that is, a steady state problem [9].

4. SOLUTION METHODS BASED ON ENTHALPY AND TEMPERATURE

Let us now turn our attention to the enthalpy based methods, also called "single-region methods" and "weak formulations." These methods are based on the fact that when two phases are present, temperature gradients cause heat conduction, which has the primary effect of changing the enthalpy which, in turn, changes the temperature. The energy balance is between rate of heat conduction and rate of change of enthalpy, not rate of change of temperature. The commonly encountered mathematical formulation of the enthalpy model is a differential one which, however, is invalid at the interface for a pure substance, because i and k ∇ T change discontinuously across the interface. The proper formulation is an integral one:

$$\frac{d}{dt} \int_V \rho i dV + \int_A \rho i \vec{v} \cdot \vec{dA} = \int_A k \nabla T \cdot \vec{dA} \qquad (4.1)$$

Note that no assumption is made that the density is constant, and the second term of (4.1) accounts for convective heat transfer.

It can be shown [16] that the conservation equations (2.1)-(2.3) are derivable from the enthalpy equation (4.1). Thus, the two formulations are equivalent.

In the second term of equation (4.1), the velocity of the liquid is needed. In order to calculate the temperature field, it would be necessary to solve flow equations in the liquid region. This, however, is a task that one would like to avoid, because of the difficulties involved. Fortunately, there are two special cases, which pertain to many practical problems, for which the velocity field need not be computed. In the first case, the influence of density change is assumed to be negligible, and heat transfer in the liquid is assumed to be by conduction only, so that $\vec{v} = 0$ everywhere. In the second case, the solid and liquid are recognized to be different in density, but the liquid is assumed to be at the saturation temperature initially. Here, if freezing is induced by removal of heat at an external boundary, the liquid remains at T_{sat} until it solidifies. For the two special cases we have just described, equation (4.1) reduces (see

[16]) to

$$\frac{d}{dt} \int_V \rho(i-i_\ell)\, dV \quad = \int_A k \;\; \nabla T \cdot \overrightarrow{dA} \tag{4.2}$$

Equation (4.2) is the basic equation of the enthalpy model and the solution methods based thereupon. In its derivation, nothing has been said about the phase change characteristics of the substance. In fact, equation (4.2) applies to phase change of substances that melt over a finite range of temperatures, and also to problems involving no phase change at all. Herein lies the advantage of the enthalpy model. No track of the interface is kept, and computer programs can be written based on it without assuming anything about the nature of the substance. For obtaining the solution, of course, a specification of the properties of the substance will be necessary, but this supplementary information can be incorporated into a small subroutine independently of the main program. Specifically, the subroutine should return the temperature corresponding to a given value of enthalpy.

For a substance that changes phase over a range of temperatures, both i and ∇T are continuous with T, and (4.2) can be reduced to a differential equation. Here, there is no clearly defined moving boundary.

Note that, in the enthalpy model, there are two dependent variables, temperature and enthalpy. For substances that change phase over a range of temperatures, enthalpy is a single valued function of temperature, and one may work with either i or T as the primary dependent variable. For a pure substance, however, there is a range of enthalpies, $i_s < i < i_\ell$, for which $T = T_{sat}$. Therefore, enthalpy must be the primary variable. Several investigators have approximated a pure substance by one which changes phase over a narrow range of temperatures, so that temperature may continue to be used as the primary variable. We shall show later that such an approximation may not be acceptable.

Next, let us see how discrete versions of the enthalpy equation are obtained, considering finite differences first. To obtain a nodal equation, we apply equation (4.2) to the element control volume surrounding the node, assigning the volume-averaged enthalpy of the element to the node; the nodal temperature is to be calculated from the nodal enthalpy using the i-T relation for the substance. For elements that contain a single phase of a substance with a discrete melting point, and for any element containing a substance with a diffuse melting range, the nodal equations that result are the same, and derived in the same way, as those for a transient problem with no phase change. That leaves for treatment the two-phase element with a pure substance. For this element, we set the nodal temperature equal to T_{sat}. In the liquid part of the element, the product ρi is replaced by the saturation value $\rho_\ell i_\ell$. Similarly, in the solid part, ρi is replaced by $\rho_s i_s$. The resulting enthalpy equation for the element can be shown (see [16]) to be

$$\rho_s V \frac{di_{node}}{dt} = \int_{A_{element}} k \nabla T \cdot \vec{dA} \tag{4.3}$$

The nodal enthalpy equation (4.3) is now similar to that for single-phase elements. For a solid element, we get equation (4.3) identically whereas for a liquid element we get equation (4.3) with ρ_s replaced by ρ_ℓ.

The obtaining of finite difference equations from equation (4.3) now follows the same procedure as for an ordinary heat diffusion problem. Equation (4.3) is deceptively simple. It accounts for phase change even in the presence of density change. If the nodal enthalpy were defined in any way other than as a volume average, the resulting equation would be more complicated.

Finite element methods have not been developed for treating problems involving discrete phase change. It is not easy to choose basis functions that are sufficiently smooth in single-phase regions and at the same time are capable of accommodating the discontinuous changes of i and k ∇T at the interface. This is an area needing more research effort.

We shall now survey the existing work based on the enthalpy model, restricting ourselves to only those substances that have a discrete melting point, i.e., those resulting in moving boundary problems. In most of the papers the phase change is assumed, for computational purposes only, to occur diffusely over a small range of temperatures surrounding the true phase change temperature. Having employed this approximation, Hashemi and Sliepcevich [7] used the ADI scheme to solve a two-dimensional problem. As these authors account for the latent heat by using di/dT in place of ρc in the heat equation, their results are very sensitive to the magnitude of the range assumed; furthermore, the error in the result increases as the span of the range is reduced, clearly indicating nonconvergence. Meyer [12] developed a fully implict scheme with the feature that the computational effort is independent of the range except that the range should be nonzero; the scheme is unconditionally stable. Shamsundar and Sparrow [17] showed how the assumption of a finite range could be done away with, and modified the implicit scheme of Meyer [12] appropriately. The standard explicit scheme was applied by Soliman and Fakhroo and the usual problem of instability examined. Similar methods have been employed by several others, and many thermal analysis package programs contain routines based on the enthalpy model, although the descriptions of the techniques are often vague and imprecise. Comini et al [3] and Friedman [6] applied the enthalpy model with a diffuse range in conjunction with a finite element method. Their results not only exhibit a strong dependence on the range assumed, but numerical instabilities become severe as the range is reduced.

Since many techniques depend on approximating a discrete melting point by a diffuse range of temperatures, it is highly desirable to assess the errors inherent in this approximation. We shall therefore, take a simple one-dimensional problem as a test case. This also provides an opportunity of exhibiting some of the characteristics of the discretized enthalpy model. The problem is the solidification of pure saturated liquid by convective cooling at a plane wall. The correct solution to this problem was obtained independently with other numerical methods, and is shown by full lines in Figure 1, which includes an inset sketch of the problem. This abscissa variable is the nondimensional time variable τ defined in Section 3. Note that, in the absence of a characteristic length, L has been replaced by k_s/h, the ratio of thermal conductivity to convective coefficient. The ordinate to the right is the nondimensional interface position, hx^*/k. The ordinate on the left is the dimensionless wall heat flux, $\hat{q} = [h(T_w - T_\infty)]/[h(T_{sat} - T_\infty)]$; the ordinate also yields the wall temperature.

The circles, squares and diamonds show the results obtained with different spatial step-sizes. An implicit difference scheme was used, and the timewise step was successively reduced until it had no observable (to three significant digits) effect. For all three spatial step-sizes, the prediction of the interface is seen to be excellent. The heat flux predicted, on the other hand, is wavy in nature, and the waviness is directly influenced by the spatial step-size. This behavior is characteristic of the enthalpy method, and has been observed by others [8]; it is caused by holding the temperature of a two-phase node constant at T_{sat}. When Ste is small, the transient proceeds slowly and the temperature profiles are quasisteady in nature. As an element solidifies, $\hat{q}$ is nearly constant; when the element becomes completely solid, its temperature is allowed to change, and the heat flux drops rapidly to its new quasisteady value. (This waviness also afflicts the interface position in multidimensional problems.) Fortunately, as the circles show, the waviness diminishes and the predicted values converge to the correct solution as the spatial step-size is reduced.

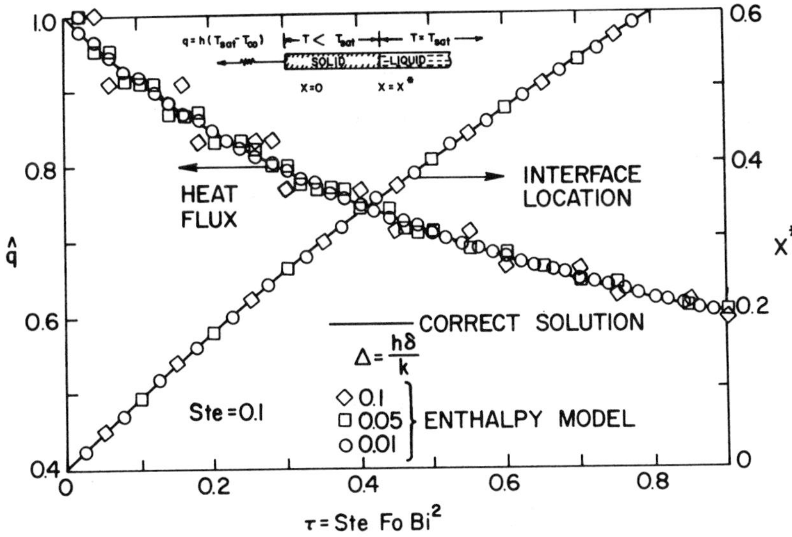

Figure 1. Effect of Spatial Step Size on Results Obtained with Discretized Enthalpy Model

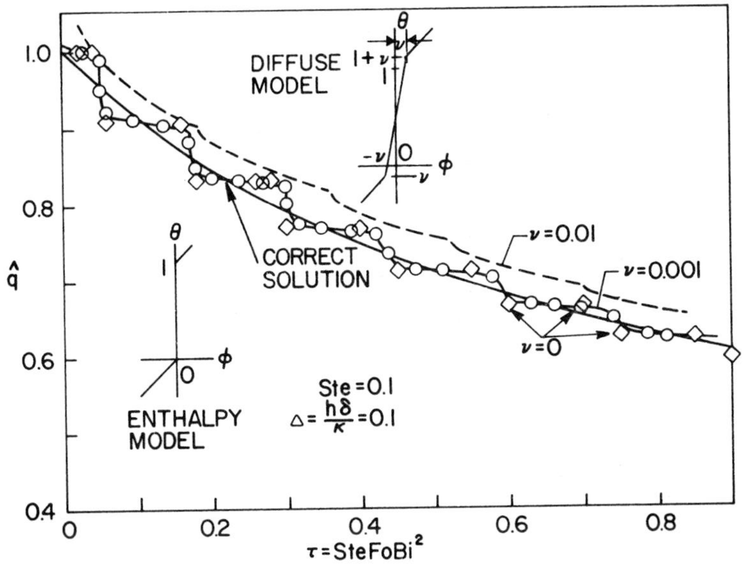

Figure 2. Comparison of Heat Flux Predictions of Enthalpy Model and Diffuse Model

It should also be pointed out that the results show how much more difficult it is to calculate fluxes and temperatures, rather than the interface position, accurately. Too often, the power of a method has been tied to its ability to predict the interface location. In many practical applications, fluxes are of greater interest than the interface.

Next, let us examine how a diffuse approximation predicts the same results. If we define $\Theta = (i - i_s)/\lambda$, $\phi = c_s\ (T-T_{sat})/\lambda$ as the dimensionless enthalpy and temperature variables, then the relation between θ and ϕ is assumed to be as shown in the inset at the top center of Figure 2. (The other inset shows the same for a pure substance.) While this is not the only diffuse approximation possible, it has the feature of coinciding with the discrete model in the single-phase regions for $|\phi| > \nu$. The figure shows heat flux curves only. (There is a dilemma as to the value of ϕ on the interface, so that no interface positions were computed. In fact, there is more than one two-phase element.) When the range parameter ν is large, the $\hat{q}$ - T curve is relatively smooth; this is desirable, but the whole curve is displaced from the correct result. As ν is reduced, the predicted curve draws near to the correct one, but the waviness increases. In fact, the results for $\nu = 0.001$, shown by circles and dashes, almost coincide with those for the discrete enthalpy model (ν = 0, shown by diamonds).

Thus, we have shown that the diffuse model is not a good approximation. The discrete enthalpy model is just as easy to use, is truer to physical reality, and easily yields the interface position. Its only undesirable feature, namely, the waviness, may be overcome by employing a sufficiently fine spatial grid. On the other hand, finite element discretizations exist only for the diffuse model. The foregoing discussion shows that the results obtained by means of the diffuse approximation should be used with caution for a pure substance.

Recently, a novel technique has been proposed by Berger et al [1]. From the enthalpy distribution i_t at time t, the enthalpy distribution $i_{t+\Delta t}$ at time $t + \Delta t$ is calculated as follows. From the enthalpy distribution at time t, the temperature distribution T_t is calculated. Using this temperature distribution as the initial one, the ordinary heat equation without phase change is solved, by any convenient technique, to get the temperature $T_{t+\Delta\tau}$ at time $t+\Delta t$. From these temperatures, the enthalpies are calculated by setting $i_{t+\Delta t} = i_t + \rho c\,(T_{t+\Delta t} - T_t)$. An advantage of the method is that the matrix of the simultaneous equations arising from discretization is invariant as phase change proceeds. Therefore, only one inversion is required; advancing the solution in time involves only matrix multiplication. It can be shown that when the standard explict difference scheme is used this technique is identical to the basic enthalpy method. However, when an implicit scheme is used, meaningful results will be obtained only when the timewise step is small enough. This is because the technique violates physical reality in assuming that heat conduction directly causes temperature change, rather than enthalpy change. When Δt is small, the error, which is not cumulative, becomes tolerable. In conclusion, there is a trade-off between the ability to use a preinverted matrix and the need for small Δt.

5. COMPARISON OF METHODS AND HINTS FOR SELECTION

At the outset, we may compare between the methods discussed in Section 3 with those of Section 4. For a given mesh, the temperature based methods, especially the ones based on immobilization, will yield more accurate results than those based on enthalpy. Therefore, the former will be more economical with respect to computer time. On the other hand, the enthalpy based methods are very much easier to program, and existing computer packages for thermal analysis may be extended to calculate phase change with only slight modifications. Although the enthalpy methods keep no track of the interface, the interface location may always be extracted from the enthalpy distribution with little effort.

Among the temperature based methods, immobilization techniques need some analytical work before programming, and extra computations are necessary for the return to physical space after obtaining the solution. Immobilization techniques and the imbedding methods have certain limitations on the shape the interface can assume for their success.

Among discretization methods, the explicit difference scheme is very convenient but, because of instability problems, it is useful only for phase change situations involving a large Stefan number and where the total time taken by the transient is small. For other situations, implicit methods must be resorted to, and systems of nonlinear algebraic equations will need to be solved. Finite element techniques possess many advantages over finite difference techniques such as easy and accurate handling of irregular boundaries, etc., but so far they have been developed only for phase change over a range of temperatures.

6. AREAS FOR RESEARCH

The following topics involve unsolved problems of technological interest; therefore, research on them would be very worthwhile.

- Finite element techniques for discrete phase change.
- Stable explicit difference methods suitable for phase change.
- Mathematical models to account for natural convection effects on melting.
- Physical models to predict cavity formation as a result of density difference.
- Physical models to account for nonequilibrium effects, supercooling, dendritic growth, etc.

NOMENCLATURE

Bi	Biot number
c	Specific heat
Fo	Fourier number
h	Convective coeficient
i	Enthalpy variable
i_ℓ, i_s	Enthalpies of liquid and solid at T_{sat}
k	Thermal conductivity
L	Characteristic length
n	Coordinate normal to interface
$\vec{r}$	Position vector
r	Radial coordinate
Ste	Stefan number
t	Time
T	Temperature variable
T_{sat}	Saturation temperature
$\vec{v}$	Velocity of liquid
v_Σ	Velocity of interface along n
x*	Coordinate of interface
α	Thermal diffusivity
δ, Δ	Spatial step size
$\Delta\tau$	Step size in τ
Θ	Nondimensional temperature
λ	Latent heat
ρ	Density
τ	Nondimensional time
Subscript s	Solid
Subscript ℓ	Liquid
Superscript '	Dimensional coordinate

RERERENCES

1. A. E. Berger, J.C.W. Rogers, H. Brezis, M. Ciment, paper presented in the Poster Session of this symposium.

2. J.A. Bilenas, L.M. Jiji, Proc. IV Int. Heat Transfer Conf., Paris-Versailles, 1, 1970, Paper No. Cu. 2.1.

3. G. Comini, S. del Guidice, R.W. Lewis, O.C. Zienkiewicz, Int. J. Num. Meth. Eng., 8, 1976, pp. 613-624.

4. J. Crank, R.S. Gupta, Int. J. Heat Mass Transfer, 18, 1975, pp. 1101-1107.

5. J.L. Duda, M.F. Malone, R.H. Notter, J.S. Vrentas, Int. J. Heat Mass Transfer, 18, 1975, pp. 901-910.

6. E. Friedman, Proc. 17th Nat. Heat Transfer Conf., 1977, pp. 182-187.

7. H.T. Hashemi, C.M. Sliepcevich, Chem. Eng. Prog. Symp. Ser. No. 79, 63, 1967, pp. 34-41.

8. K. Katayama, M. Hattori, Bull. JSME, 18, 1975, pp. 41-46.

9. P.G. Kroeger, S. Ostrach, Int. J. Heat Mass Transfer, 17, 1974, pp. 1191-1207.

10. A. Lazaridis, Int. J. Heat Mass Transfer, 13, 1970, pp. 1459-1477.

11. G.H. Meyer, Int. J. Num. Meth. Eng., 11, 1977, pp. 741-752.

12. G.H. Meyer, SIAM J. Num.Anal., 10, 1973, pp. 522-538.

13. W.D. Murray, F. Landis, Trans. ASME, J. Heat Transfer, 81, 1959, pp. 106-112.

14. K.A. Rathjen, L.M. Jiji, Trans. ASME, J. Heat Transfer, 93, 1971, pp. 101-109.

15. T. Saitoh, Trans. ASME, J. Heat Transfer, 99, 1977 (to be published).

16. N. Shamsundar, E.M. Sparrow, Trans. ASME, J. Heat Transfer, 98, 1976, pp. 550-557.

17. N. Shamsundar, E.M. Sparrow, Trans. ASME, J. Heat Transfer, 97, 1975, pp. 333-340.

18. R. Siegel, M.E. Goldstein, J.M. Savino, Proc. IV Int. Heat Transfer Conf., Paris-Versailles, 1, 1970, Paper No. Cu. 2.11.

19. F.W. Spaid, A.F. Charwat, L.G. Redekopp, R. Rosen, Int. J.

Heat Mass Transfer, 14, 1971, pp. 673-687.

20. G.S. Springer, D.R. Olson, ASME Paper No. 62-WA-246, 1962.

21. L.C. Tien, J.O. Wilkes, Proc. IV Int. Heat Transfer Conf., Paris-Versailles, 1, 1970, Paper No. Cu. 2.12.

ACKNOWLEDGEMENTS

This work was funded in part by the U.S. Energy Research and Development Administration, Contract No. EG-77-C-04-3974, Subcontract No. EFT-5.

N. Shamsundar
Department of Mechanical Engineering
University of Houston
4800 Calhoun
Houston, Texas 77004.

THE APPLICABILITY AND EXTENDABILITY OF MEGERLIN'S METHOD FOR SOLVING PARABOLIC FREE BOUNDARY PROBLEMS

A. D. Solomon

We examine the applicability of an analytical approximation procedure of F. Megerlin to a number of free boundary problems; we then indicate a method for extending the Megerlin procedure to a convergent approximation technique, prove its convergence for a particular family of problems, and exhibit its implementation.

1. INTRODUCTION

In [9] F. Megerlin introduced a method for the approximate solution of melting and soldification problems. Reminiscent of [14], the method is based on the substitution of a suitable expression for the temperature into the conditions of the problem and the solution of a resulting differential equation for the location of the free boundary. The Megerlin approach has been applied to a number of phase change problems ([4], [17], [18], [22]) although no error analysis or convergent extension has been developed for it. The virtue of the method is that it yields simple but relatively accurate approximations to the phase front and the temperature distribution and can thereby furnish important information about the problem without extensive computation.

Upon examination, one finds that the Megerlin method is applicable to a much wider variety of free boundary problems than those encountered in simple phase change processes; in addition, by means of a finite-element type approach it can be extended to a method which hopefully is convergent to the exact solution of the free boundary problem to which it is applied. These are the subjects of the present paper. We begin by exhibiting the applicability of the Megerlin procedure to a variety of problems. We then extend the method and apply this extension to two particular

ISBN 0-12-757350-X

Stefan problems.

2. THE APPLICABILITY OF MEGERLIN'S METHOD

In this part we apply the Megerlin method to a variety of free boundary problems. We begin with a problem for which the heat flux at the boundary is a given constant. We then turn to a problem arising in sequential decision theory having variable "critical temperature" and an initial singularity. Finally, we present two problems for which the Megerlin method yields the exact solution. We note that others are amenable to it [19].

Problem 1. A Flux-Boundary Problem:

$$u_\tau = u_{\xi\xi}, \quad 0 < \xi < \Sigma(\tau) \tag{1a}$$

$$u_\xi(0,\tau) = -1 \tag{1b}$$

$$\Sigma'(\tau) = -u_\xi\ (\Sigma(\tau),\tau) \tag{1c}$$

$$u(\Sigma(\tau),\tau) = 0 \tag{1d}$$

$$\Sigma(0) = 0 \tag{1e}$$

$$u(\xi,\tau) = 0, \quad \xi > \Sigma(\tau). \tag{1f}$$

Following [9] we represent $u(\xi,\tau)$ in the form

$$u(\xi,\tau) = \sum_{j=1}^{2} a_j(\tau)\ (\xi - \Sigma\ (\tau))^j. \tag{2}$$

Whereupon by (1a,c)

$$u = -\Sigma'\ (\xi-\Sigma) + \frac{\Sigma'^2}{2}\ (\xi-\Sigma)^2. \tag{3}$$

Substitution into (1b) produces an ordinary differential equation which, when solved, yields the relation

$$\tau = \frac{1}{2}\{\Sigma + \frac{1}{6}(1 + 4\Sigma)^{3/2}\} - \frac{1}{12} \tag{4}$$

In [15] M. Rose has computed the solution to Problem 1 using a weak solution approach. A comparison of his results with (4) is shown in Table I.

Table I. $\tau = \tau(\Sigma)$ from [15] and (4)

Σ	τ ([15])	τ (eqn. 4)
0	0	0
.2	.220	.218
.4	.450	.466
.6	.730	.739
.8	1.030	1.034
1.0	1.350	1.348
1.2	1.690	1.681
1.4	2.050	2.030
1.6	2.440	2.394

These values agree to within the limits of accuracy of [15].

Problem 2. A Problem in Sequential Decision Theory:

Our second problem arises in work by Chernoff and others ([1], [10], [16]) in statistical decision theory. This problem exhibits singular behavior at the free boundary and takes the following form.

$$u_\tau = u_{\xi\xi}, \quad 0 < \xi < \Sigma(\tau) \tag{5a}$$

$$u_\xi(0,\tau) = \frac{1}{2} \tag{5b}$$

$$u_\xi(\Sigma(\tau),\tau) = 0 \tag{5c}$$

$$u(\Sigma(\tau),\tau) = \frac{1}{2\tau} \tag{5d}$$

$$\Sigma(0) = 0 \tag{5e}$$

Let us seek a representation of u in the form

$$u = \frac{1}{2\tau} + \sum_{j=1}^{3} a_j(\tau)\ (\xi - \Sigma(\tau))^j \tag{6}$$

Substitution implies

$$u(\xi,\tau) = \frac{1}{2\tau} - \frac{1}{4\tau^2}(\xi - \Sigma(\tau))^2 + \frac{\Sigma'(\tau)}{12\tau^2}(\xi - \Sigma(\tau))^3 \tag{7}$$

whence by (5b)

$$\Sigma'(\tau)\ \Sigma(\tau)^2 + 2\Sigma(\tau) = 2\tau^2 \tag{8}$$

Let us seek a solution in the form

$$\Sigma(\tau) = \tau^2 + \varepsilon \tag{9a}$$

for ε', ε^2 small compared to ε. Substituting (9a) into (8) and ignoring all terms but ε yields

$$\varepsilon = \frac{-\tau^5}{1 + 2\tau^3} \tag{9b}$$

whence from (9a),

$$\Sigma(\tau) \doteq \tau^2\ \frac{1 + \tau^3}{1 + 2\tau^3} \tag{10}$$

In Table II we compare (10) with the results of [10], the latter obtained by a method of lines approach. We see that (10) is an effective approximation to $\Sigma(\tau)$ over the range indicated, whose relative error varies at most up to about 12%.

Table II. $\Sigma(\tau)$ for Problem 2

τ	$\Sigma(\tau)$ ([10])	$\Sigma(\tau)$ (10)	Relative Error $= \frac{([10]) - (10)}{([10])}$
0	0	0	0
.05	.0025	.0025	0
.1	.0100	.0100	0
.2	.0398	.0397	.0025
.3	.0886	.0877	.0102
.4	.1543	.1509	.0220
.5	.2343	.2250	.0397
.8	.5317	.4781	.1008
1.0	.7565	.6666	.1188
1.4	1.2289	1.1310	.0797
1.8	1.7059	1.8977	-.1124
2.0	1.9420	2.1176	-.0904

Problem 3. A Non-Standard Problem:

The following problem (of [11]) differs from the preceding in that there are heat source terms in the heat equation and at the free boundary, while the boundary condition varies.

Find u, Σ such that

$$u_{\xi\xi} - u_{\tau} = f(\xi,\tau) \tag{11}$$

$$u(0,\tau) = g(\tau) \tag{12a}$$

$$u(\Sigma(\tau),\tau) = 0 \tag{12b}$$

$$\Sigma'(\tau) + u_{\xi}(\Sigma(\tau),\tau) = h(\Sigma(\tau),\tau) \tag{12c}$$

$$\Sigma(0) = 0. \tag{12d}$$

As is easily seen, for the sets of functions f, g, h chosen in [11] the Megerlin method yields the exact solution.

3. SOME ADDITIONAL REMARKS

The preceding examples show that the Megerlin approach may be very accurate, both qualitatively and quantitatively, for short times. For longer times in the absence of additional independent information about the solution, one may be justifiably hesitant about using the method and a simple finite difference check might be called for, via e.g. [21]. We note that corrective methods have been applied to the integral profile method of Goodman [4] and to the Megerlin method [9], the latter technique based on energy conservation.

We now turn to an extension of the Megerlin method by means of which it may be possible to extend its range of applicability.

4. THE EXTENDABILITY OF THE MEGERLIN METHOD

We now discuss an extension of Megerlin's method which is based on a domain subdivision and the application of some variant of the Megerlin procedure to each of the resulting subregions. The approach is in the spirit of finite-element techniques and has been broached in the past in [12], [13].

The method is described by applying it to two simple Stefan problems. The first problem considered is for constant boundary and initial conditions on a simi-infinite slab. We then apply the procedure to Problem 1 over a long time interval.

Consider the following problem.

Problem 4. -- Find X, T such that

$$T_t = \alpha T_{xx}, \quad 0 < x < X(t) \tag{13a}$$

$$T(0,t) = T_0 > T_{cr}, \quad t > 0 \tag{13b}$$

$$T(X(t),t) = T_{cr}, \quad t > 0 \tag{13c}$$

$$-KT_x(X(t),t) = \rho HX'(t), \quad t > 0 \tag{13d}$$

$$X(0) = 0 \tag{13e}$$

Here α, K, c, ρ, H, T_{cr} represent the diffusivity, conductivity, specific heat, density, latent heat and melt temperature. From [2] the solution to Problem 4 is given by

$$X(t) = 2\lambda\sqrt{\alpha t} \tag{14a}$$

$$T(x,t) = T_0 - \frac{(T_0 - T_{cr})}{\operatorname{erf}\lambda}\operatorname{erf}(x/2\sqrt{\alpha t}) \tag{14b}$$

where λ is the unique root of

$$\lambda e^{\lambda^2}\operatorname{erf}\lambda = \frac{c(T_0 - T_{cr})}{H\sqrt{\pi}} \tag{14c}$$

and

$$\operatorname{erf} z = \frac{2}{\sqrt{\pi}}\int_0^z e^{-s^2}\,ds \tag{14d}$$

We now define the extended Megerlin method. Let N = 1,2, ... by any natural number. For j = 0, 1, 2, ..., N let

$$X_j(t) = \frac{j}{N}X(t), \; j = 0, 1, \ldots, N. \tag{15}$$

On the <u>jth</u> subdivision

$$X_{j-1} < x < X_j, \; j = 1,N$$

let

$$T^j(x,t) = \Gamma_j(t) + a_j(t)(x - X_j) + b_j(t)(x - X_j)^2. \tag{16}$$

We will require that T^j obey the heat equation (13a) at $x = X_j$ and that, together with its first order x-derivative, it pass continuously to the values of T^{j+1}, T_x^{j+1} across $x = X_j$. Moreover T^1 and T^N are to obey (13b,d) respectively.

After some manipulation we are led to the system of equations

$$a_j = a_{j+1} - \frac{2b_{j+1}X}{N}, \quad j = 1, N \tag{17a}$$

$$\rho H X' = -Ka_N \tag{17b}$$

$$T_0 = \Gamma_1 - \frac{a_1 X}{N} + \frac{b_1 X^2}{N^2} \tag{17c}$$

$$\Gamma_N = T_{cr} \tag{17d}$$

$$\Gamma'_j + \frac{jKa_j a_N}{jHN} = 2\alpha b_j, \quad j = 1, N \tag{17e}$$

$$a_{j+1} + a_j = \frac{2N}{X}(\Gamma_{j+1} - \Gamma_j), \quad j = 1, N - 1 \tag{17f}$$

$$a_N = \frac{HN}{cX}\left\{1 - \sqrt{1 + \frac{2c}{H}(\Gamma_N - \Gamma_{N-1})}\right\} \tag{17g}$$

Let us search for solutions a_j, b_j, Γ_j, X of the form

$$\Gamma_j = \text{const. for all } j \tag{18a}$$

$$X = 2\lambda\sqrt{\alpha t} \tag{18b}$$

Substitution yields the following transcendental equation for a root λ which depends on N.

$$\frac{c}{H}(T_0 - T_{cr}) = \frac{\lambda^2}{N}\left\{1 + 3\left[1 + \frac{2\lambda^2}{3N^2}\right]\prod_{K=2}^{N}\left\{1 + \frac{2K\lambda^2}{N^2}\right\}\right.$$

$$\left. + 2\sum_{K=2}^{N-1}\prod_{\ell=K+1}^{N}\left\{1 + \frac{2\ell\lambda^2}{N^2}\right\}\right\} \tag{19}$$

For each N the right-hand side of (19) is a monotonically increasing function of λ, vanishing for $\lambda = 0$. Thus (19) has a unique root λ dependent on N and $c(T_0 - T_{cr})/H$. Moreover, by a short calculation

$$\lambda < \left\{\frac{c(T_0 - T_{cr})}{2(1 - \frac{2}{N})H}\right\}^{\frac{1}{2}}$$

whence, λ is bounded by a number independent of N.
It is easily seen that (19) is of the form

$$\frac{c(T_0 - T_{cr})}{H\sqrt{\pi}} = 0(\frac{1}{N}) + \lambda e^{\lambda^2} \text{ erf } \lambda;$$

which, as $N \to \infty$ yields (14c) for the root λ. A similar computation yields the convergence of the Γ_j and the intermediate approximations of T to the true solution.

It is interesting to compare equations (19) and (14c). These relations may be rewritten as

$$F(\lambda) = \lambda e^{\lambda^2} \text{ erf } \lambda = \frac{c(T_0 - T_{cr})}{H\sqrt{\pi}}$$

and

$$F_N(\lambda) = \frac{c(T_0 - T_{cr})}{H\sqrt{\pi}}, \quad N = 1,2,\ldots,$$

In Table 3 we list the values of F, F_1, F_2, F_3 for $0 \leq \lambda \leq 2$ at intervals of length .2. We observe the slow convergence rate and the fact that the limits of, say, 10% relative accuracy are roughly given by

$$\frac{c(T_0 - T_{cr})}{H\sqrt{\pi}} \approx \begin{matrix} 4 & \text{for } N = 1 \\ 4.6 & \text{for } N = 2 \\ 5 & \text{for } N = 3 \end{matrix}$$

Table III. Source Functions for λ

λ	F	F_1	F_2	F_3
.2	.0464	.0469	.0467	.0466
.4	.2011	.2094	.2064	.2050
.6	.5193	.5524	.5408	.5355
.8	1.126	1.1843	1.1635	1.1568
1.0	2.2907	2.2568	2.2568	2.2738
1.2	4.6106	3.9647	4.0934	4.2224
1.4	9.4648	6.5464	7.0666	7.5347
1.6	20.208	10.284	11.726	13.03
1.8	45.459	15.501	18.818	21.937
2.0	108.69	22.568	29.338	36.065

However, this range allows us a large value of $T_0 - T_{cr}$ for many materials. Thus for the melting of ice, $\frac{c}{H\sqrt{\pi}} = 0.00394/°F$, while for a typical wax (N-tetradecane), $\frac{c}{H\sqrt{\pi}} = .00285/°F$ (see [5]), whence the temperature range yielding 10% agreement in the functions F_N would be of order of 1000°F --or for all practical purposes, without limit for even N = 1. The need for caution is evident if we consider, for example, the recrystallization of iron from type α to type γ, occurring at 1670°F, for which $\frac{c}{H\sqrt{\pi}} =$ .03808°F, [6]. Now an allowable temperature range would be of the order of 100°F for N = 1, and 130°F for N = 3, and so use of the refinement technique may be called for (although the high slope of F will certainly mollify this effect).

We now consider the problem of constant boundary flux (Problem 1). Define $\Sigma_j = j\Sigma/N$ for N a natural number and $j = 1,2,\ldots,N$ and let

$$u^j = \Gamma_j + a_j(\xi - \Sigma_j) + b_j(\xi - \Sigma_j)^2, \quad j = 1, N.$$

Then, as earlier, substitution into (1b-f), together with continuity of the u and u_x values, yields a set of relations which is augmented by (1a) in the form

$$u^j_\tau = \tfrac{1}{2}(u^{j+1}_{\xi\xi} + u^j_{\xi\xi}), \quad \xi = \Sigma_j$$

Starting from $\tau = 0$ until τ attains some appropriate "switching time" $\tau = \tau_{sw}$, we will use (4). At $\tau = \tau_{sw}$ we calculate the values $\Gamma_1(\tau_{sw}), \ldots, \Gamma_N(\tau_{sw})$ and $\Sigma(\tau_{sw})$ from (3) and hence the a_j, b_j from our scheme. We now advance Γ_j, Σ to a new $\tau > \tau_{sw}$ using our scheme. In practice this approach is numerically stable [20].

In this way we obtain a scheme for solving Problem 1. This was done on a time interval $0 \leqslant \tau \leqslant 28$ for various N values as shown in Table IV. In each case (4) was used to compute Σ up to a switching time $\tau = .2$, and a Runge Kutta scheme was used from then on. We may observe the accuracy of the results by means of the heat balance relation

$$I_1(\tau) = I_2(\tau)$$

for

$$I_1(\tau) = \tau - \Sigma(\tau)$$

(the energy not stored as latent heat) and

$$I_2(\tau) = \int_0^{\Sigma(\tau)} u(\xi,\tau)d\xi$$

(the energy stored as sensible heat).

Table IV. $\Sigma = \Sigma(\tau)$ [(49c)]: N = 1,5

τ	N=1	N=2	N=3	N=4	N=5
0	0	0	0	0	0
4	2.404	2.244	2.194	2.280	2.241
8	4.094	3.708	3.642	3.776	3.723
12	5.549	4.928	4.855	5.021	4.961
16	6.865	6.009	5.930	6.122	6.058
20	8.085	6.995	6.913	7.125	7.058
24	9.233	7.911	7.826	8.055	7.987
28	10.324	8.772	8.685	8.929	8.860

Let

$$E = \frac{I_2 - I_1}{I_1}$$

denote the relative under-estimation of latent heat stored in the system. In Table V the computed values of E are listed for various τ, N. Additional subdivisions (N = 6, 10) reduce E to a level of .02.

Table V. $E = (I_2 - I_1)/I_1$

τ	N=1	N=2	N=3	N=4
0	0	0	0	0
4	.170	.162	-.180	-.051
8	.268	.220	-.159	-.034
12	.342	.253	-.146	-.035
16	.403	.284	-.136	-.035
20	.454	.310	-.129	-.036
24	.500	.334	-.123	-.036
28	.541	.354	-.118	-.036

In all cases the computed Σ values do not exhibit appreciable change, leading us to believe that effective accuracy in Σ is achieved already for N = 2. To justify this claim let us consider Problem 1 for large τ, Σ, ξ. Introducing the change of variables.

$$y = \frac{1}{\xi}, \quad s = \frac{1}{\tau}, \quad Y = \frac{1}{\Sigma},$$

$$u(\xi,\tau) = v(y,s)$$

transforms Problem 1 to the form

$$s^2 v_s = -y^4 v_{yy} - 2y^3 vy, \quad y > Y(s)$$

$$v(Y(s),s) = 0 \qquad s > 0$$

$$s^2 Y'(s) = v_y Y(s)^4 \qquad s > 0$$

$$y^2 v_y(y,s) \to 1 \quad \text{as } y \to \infty$$

Let us attempt to find a locally linear approximation to v(y,s) for $y \approx Y(s)$ of the form

$$v = a(y-Y)$$

Substituting this into the differential equation for y = Y and solving the resulting differential equation implies

$$\Sigma = 2\sqrt{\tau}$$

This relation conforms to the asymptotic estimate

$$\Sigma(\tau) \geqslant \frac{\tau}{2\sqrt{\tau} - 1}$$

of A. Friedman [3]. Let us compare this with the results of using the subdivision procedure for the case N = 2,3, over a long time range ($0 < \tau < 200$). In Table VI we list the values of $\Sigma = 2\sqrt{\tau}$ and Σ computed for N = 2. We observe an initially falling discrepancy which beyond $\tau \approx 140$ begins to rise again. We conjecture that the initial discrepancy is due to inaccuracy of the long-time estimate while the later discrepancy-partially corrected by increasing N, is due to growing inaccuracy of the subdivision procedure for large time. The Megerlin method (4) with no subdivision is not accurate over much of the τ range of Table V. Thus, for $\Sigma = 20$ the value of τ predicted by (4) would be $\tau = 70.667$ while for $\Sigma = 30$ we find $\tau = 125.833$ whence the predicted Σ is moving appreciably faster than is expected.

Table VI. $\Sigma = \Sigma(\tau)$ for large τ

τ	$\Sigma=2\sqrt{\tau}$	Scheme for N=2	Scheme for N=3
0	0	0	0
20	8.944	6.955	6.913
40	12.649	11.118	11.024
60	15.492	14.505	14.397
80	17.888	17.485	17.358
100	20.000	20.192	20.045
120	21.909	22.700	22.529
140	23.664	25.052	24.855
160	25.298	27.280	27.054
180	26.833	29.402	29.148
200	28.284	31.436	31.151

ACKNOWLEDGEMENT

We take this opportunity to thank Robert Meacham for programming the algorithm of Section 4.

REFERENCES

1. J. Breakwell and H. Chernoff, "Sequential Tests for the Mean of a Normal Distribution II (large t)," Ann. Math. Stat. 35, 162-73 (1964).

2. H. Carslaw and J. Jaeger, Conduction of Heat in Solids, 2nd ed., Oxford at the Clarendon Press, London, 1959.

3. A. Friedman, "Remarks on Stefan-type Free Boundary Problems for Parabolic Equations," J. of Math. and Mech. 9, 885-903 (1960).

4. J. Goodling and M. Khader, "Results of the Numerical Solution for Outward Solidification with Flux Boundary Conditions," J. Heat Trans., 307-09 (1975).

5. F. D. Hale, M. Hoover and M. O'Neill, Phase Change Materials Handbook, Lockheed Missiles and Space Co., Huntsville, 1971, Rep. No. NASA CR-61363.

6. R. Hultgren, R. Orr, D. Anderson and K. Kelley, Selected Values of Thermodynamic Properties of Metals and Alloys, John Wiley, New York, 1963.

7. J. Kern, "A Simple and Apparently Safe Solution to the Generalized Stefan Problem," Internat. J. Heat Mass Trans. 20, 467-74 (1977).

8. P. Kosky, "Heat Transfer During Liquid to Solid Phase Change," Letters in Heat Mass Trans. 2, 339-46 (1975).

9. F. Megerlin, "Geometrisch Eindimensional Warmeleitung Beim Schmelzen und Erstarren," Forsch. Ing.-Wes. 9, 40-46 (1968).

10. G. Meyer, "One-Dimensional Parabolic Free Boundary Problems," SIAM Rev. 19, 17-34 (1977).

11. G. Meyer, "On a Free Interface Problem for Linear Ordinary Differential Equations and the One-Phase Stefan Problem," Numer. Math. 16, 248-67 (1970).

12. M. Mori, "Numerical Solution of the Stefan Problem by the Finite Element Method," Mem. Numer. Math 2, 35-44 (1975).

13. B. Noble, "Heat Balance Methods in Melting Problems," Moving Boundary Problems in Heat Flow and Diffusion, Proc. of the Conference at Oxford University, March 25-27, 1974, p. 208-9, ed., J. Ockendon and W. Hodgkins, Clarendon Press, Oxford, 1975.

14. C. Pekeris and L. Slichter, "Problem of Ice Formation," J. Appl. Phys. 10, 135-37 (1939).

15. M. Rose, "A Method for Calculating Solutions of Parabolic Equations with a Free Boundary," Math. Comput. 14, 249-56 (1960).

16. G. Sackett, "Numerical Solution of a Parabolic Free Boundary Problem Arising in Statistical Decision Theory," Math. Comput. 25, 425-33 (1971).

17. T. Saitoh, "An Experimental Study of the Cylindrical and Two-dimensional Freezing of Water With Varying Wall Temperature," Technology Reports, Tohoku University 41, 61-72 (1976).

18. M. Shamsundar and E. Sparrow, "Storage of Thermal Energy by Solid-Liquid Phase-Change-Temperature Drop and Heat Flux," J. Heat Trans., 541-43 (1974).

19. J. Simmons, R. Parker and R. Howard, "Theory of Whisker Growth and Evaporation," J. Appl. Phys. 35, 2271-72 (1964).

20. G. Smith, Numerical Solution of Partial Differential Equations Oxford University Press, London, 1965.

21. A. Solomon, "Some Remarks on the Stefan Problem," Math. Comput. 20, 347-60 (1966).

22. K. Stephan, "Schmelzen und Erstarren Geometrisch Einfacher Korper," Kaltetechnik-Klimatisierung 23, 42-46 (1971).

Applications Papers

MOVING BOUNDARY PROBLEMS

AN APPLIED OVERVIEW OF MOVING BOUNDARY PROBLEMS

Bruno A. Boley

An overview of moving-boundary problems is sketched in a general way with emphasis on some types of problems in which further work appears to be desirable from an applied viewpoint. Areas touched are those of the mechanical behavior of melting or solidifying bodies, the establishment of suitable constitutive equation, the use of approximate methods, the state of multidimensional problems, and the application of continuum approaches to the study of crystal growth.

1. INTRODUCTION

Under the three headings of Theory, Methods and Applications, the present conference spans the remarkably broad body of current research in problems involving moving boundaries. A glance at the program will convince anyone of the impossibility of summarizing in a single paper even the principal features of the work being done in any one of the three catagories, without degenerating into a tedious, although possibly useful, extensive bibliography. Moreover, an attempt at such a summary at the beginning of the conference cannot possibly include adequate reference to the work to be presented during the next succeeding days, and thus *ipso facto* must fall short of an up-to date description of the state of the field.

The panel discussion on the last day of the conference will nevertheless endeavor to formulate some general observations on the state of current knowledge and current needs. I thought it would be more appropriate at this time to indicate certain areas of research which, in my opinion, have a distinct practical importance, but are nevertheless insufficiently emphasized or absent at the present conference. The principal such areas are the following:

ISBN 0-12-757350-X

- -Studies of the mechanical response accompanying changes of phase
- -Determinations of material behavior and constitutive relations in regimes encountered in change of phase problems
- -The use of approximations and of approximate methods
- -Analytical work on multidimensional problems
- -Correlations between approaches at the microscopic ("material science") and continuum("mechanics") scales

Each of these will be discussed in what follows, together with some general observations which may on occasion be appropriately introduced. In all cases, no effort will be made at a comprehensive bibliography, and only typical researches will be indicated; these and the references cited therein should nevertheless form a good introduction to the work in any one of the areas under discussion. In addition, it may be well to mention here an earlier conference [1.1] which also brought together--to the author's knowledge, for the first time--mathematicians and practicing engineers active in this field.

2. MECHANICAL BEHAVIOR DURING CHANGES OF PHASE

The deformations and stresses accompanying problems of change of phase are of great importance in a variety of applications. As examples one might mention the stresses arising during casting solidification [2.10,18] the deformations responsible for air-gap formation between casting skin and mold [2.12,14,17]and the accompanying phenomenon of interface instability[2.11,19],the deformation ablating shields [2.9,15] and solid propellants [2.13] and those of nuclear reactor fuel elements [2.5].

The first question that must be raised with regard to problems of this kind concerns the permissibility of uncoupling the thermal and mechanical effects, as indeed was done in all the works mentioned above. Such uncoupling is permissible in ordinary thermoelastic problems, when the coupling parameter.

$$\delta = \frac{(3\lambda + 2\mu)^2\alpha^2T^*}{(\lambda + 2\mu)\rho c_v} \tag{2.1}$$

is small compared to unity, and dynamic effects are small [2.7]. In this equation λ and μ are the Lamé elastic constants, α is the coefficient of linear expansion, ρ the density, c_v the specific heat at constant colume and T^* a reference (absolute) temperature at which the material is stress-free. The latter effect depends principally on the magnitude of a thermal velocity (speed of interface advance, or of isotherm motion) relative to the speed of stress waves in the solid; this is normally sufficiently small to cause mechanical effects to be passed (other than after reflections) by the time the thermal ones become significant. An elastic analysis of the dynamic stresses accompanying a melting process may be found in [2.16]; the stresses in the Neumann problem in the direction of solidification may be the same order of magnitude as those arising in the other directions under static conditions [cf.eq.(2.6)], but can be expected to decrease rapidly if the actual temperature does not experience instanteneous jumps (cf. 2.7]).

An estimate of the effect of δ may be obtained by recalling [2.14] that in simple thermoelastic problems it is equivalent to a replacement of the diffusivity $\varkappa$ by the effective diffusivity

$$\varkappa^* = \frac{\varkappa}{1 + \delta} \tag{2.2}$$

and by noting that this statement carries over to moving-boundary solutions of the Neumann type. This is best seen by regarding such solutions as arising from linear combinations of elementary heat-conduction solutions for a half-space with arbitrary conditions at the surface and at the far end, and initially(cf. [2.4]); then use of the effective diffusivity $\varkappa^*$ applies to each elementary solution, provided that in each the surface tractions are required to vanish. This follows from integration of the one-dimensional equation of equilibrium in the absence of inertia effects, namely

$$(\lambda + 2\mu)u_{x,xx} - (3\lambda + 2\mu)\alpha T_{,x} (\equiv \sigma_{xx,x}) = 0 \tag{2.3}$$

where u_x and σ_{xx} represent displacement and stress components respectively, and commas indicate differentiation. Substitution into the linearized coupled heat equation.

$$KT_{,xx} = \rho c_v \dot{T} + (3\lambda + 2\mu)\alpha T^* \dot{u}_{x,x} \tag{2.4}$$

(where K is the thermal conductivity and dots indicate differentiation with respect to time) gives

$$\overset{*}{\kappa} T_{,xx} = \dot{T} \tag{2.5}$$

The stress component σ_{xx} is, as has been noted, identically zero, and so are, of course, the shears. The only non-zero stress components are (in the melting case)

$$\sigma_{yy} = \sigma_{zz} = -\frac{\alpha ET}{1 - \nu} \tag{2.6}$$

where E is Young's modulus and ν is Poisson's ratio.

The above results have been obtained for the purely elastic one-dimensional Neumann problem, but may be practically useful in providing a preliminary estimate of the magnitude of this effect even in more complicated problems where its neglect may be questioned. In any case, certain parts of the solution (e.g., the coefficient λ in the Neumann expression for the interface positions, i.e.,

$$s(t) = 2\lambda\sqrt{\kappa t} \tag{2.7}$$

(where t is time) are not likely to be very sensitive to changes in diffusivity; λ in fact is independent of κ for equal properties in the solid and liquid [2.8,11]. Extension of these results to problems other than the ones for which they are strictly valid may be exemplified by an examination of (still one-dimensional) cases in which the moduli are temperature dependent, in which (with $\sigma_{xx} \equiv 0$ and $\sigma_{yy} = \sigma_{zz} \equiv \sigma$)

$$u_{x,x} = f_1(\sigma,T) \; ; \; u_{y,y} = u_{z,z} \equiv 0 = f_2(\sigma,T) \tag{2.8}$$

or

$$u_{x,x} = F(T) \tag{2.8a}$$

where the effective diffusivity is

$$\varkappa^{\varkappa} = \frac{\varkappa}{1 + \delta \frac{dF}{dT}} \tag{2.8b}$$

since similarity still holds in problems with temperature-dependent properties [2.4]. The last statement is no longer valid if the solid is viscoelastic, since in that case (even on the normal assumption of elastic behavior under hydrostatic conditions [2.7])

$$(\lambda + 2\mu)\bar{u}_{x,x} = (3\lambda + 2\mu)\ f(p)\alpha\bar{T} \tag{2.9}$$

where bars indicate Laplace transforms and p is the Laplace transform parameter. Thus the transformed heat equation becomes (since dissipative effects vanish in the one-dimensional problem [2.7]):

$$\varkappa\bar{T}_{,xx} = p\bar{T}[1 + \delta f(p)] \tag{2.9a}$$

This somewhat detailed discussion of coupling should not be taken to imply that this effect is often likely to assume major importance, nor is it an open invitation to mathematicians to plunge heedlessly into a relatively unexplored field. It is rather an expression of the feelings that this very lack of experience makes it prudent to proceed with caution, and to give some thought to the validity of the inevitable underlying assumption of the physical model under study. Some rules of thumb permitting some preliminary estimates of the validity of such assumptions are therefore particularly useful.

Whether coupling is important or not, the mechanical response itself of course is. In particular, the stresses of eq. (2.6) are present regardless of coupling. In the solidification case, these stresses are

$$\sigma_{yy} = \sigma_{zz} = -\frac{\alpha\ E(T - T_m)}{1 - \nu} \tag{2.10}$$

on the assumption that no bending occurs during the formation of the solidified skin. Should such bending occur, they may at times be estimated as in [2.6]; in particular if the skin is free of bending and axial restraints, they cannot exceed $(4/3)|\alpha E\Delta T/(1-\nu)|$, where ΔT is the maximum temperature excursion in the skin. No studies of stresses and deformation in multidimensional problems have yet been made to the author's knowledge.

The calculation of the mechanical response requires, first of all, an accurate knowledge of temperature distributions, gradients, and history (some discussions of the errors which may arise even in non-moving-boundary problems because of inaccuracies in the temperature may be found in [2.3,6]). It cannot be over emphasized that analyses which are sufficiently accurate for "thermal" purposes may not be so for stress calculations, a fact that may be easily overlooked particularly when dealing with purely numerical solutions. In fact the development of some information regarding the accuracy required in the temperature for desired accuracy of stresses and deformations would seem to be a useful subject of future mathematical research.

It should be clear from the preceding discussion, however, the type of material behavior assumed in the analysis will greatly influence the results; thus, before proceeding much further, it is necessary that more data on this subject be developed experimentally.

3. MATERIAL PROPERTIES AND CONSTITUTIVE EQUATIONS

Considerable data have been gathered by various investigators on the variation of material properties with temperature, in some cases approaching the melting temperature. Much more is known regarding the thermophysical properties (e.g.,[3.1]) than mechanical ones, although many investigations for specific materials have been undertaken over the years (see, for example, the 21 studies of low-carbon steels listed in [3.3]). Information of this type is of course extremely useful, but falls somewhat short of the more general goal of establishing constitutive relations appropriate to a given material under given conditions. For example, the studies cited in [2.18],[2.17]and[2.12], treat the solidifying ingot as temperature-dependent elasto-plastic, elastic, and non-linearly viscoelastic respectively. There is some conceptual difficulty with the first of these, in that uniqueness of solution is not assured (the uniqueness theorems are incremental [2.7], and uniqueness of the initial stress state in a strip of zero initial thickness is difficult to establish; no such difficulty arises in analogous melting problems, e.g. [2.9] or [2.5]. Elastic behavior is almost certainly not to be found in reality in the regimes encountered in change-of-phase problems, but that does not mean that such studies are without value: they may provide, at the very least, an extremely useful initial understanding of the problem, and, in some instances, a useful approximation. Thus, it was noted in [2.9] that the deformations of a melting elasto-plastic plate would be well approximated (from above) by those calculated elastically. When melting occurs as a result of internal heat generation, the approximation still holds, but from below [2.5].

A combination of temperature-dependent plastic and non-linear viscoelastic behavior will undoubtedly describe most desired conditions. However, even aside from the fact that the material properties in such a model are quantitatively not known, what is needed is some knowledge of the simplest model that may properly be used. For example, a simple work-hardening behavior was found to be exhibited by the austenitic steel in [3.2], but not by the delta-ferritic iron of [3.3], the latter however lending itself to a simple relation for the stress (under conditions of constant strain), namely

$$\sigma = A(\dot{\varepsilon})^m \exp(B/T) \qquad (3.1)$$

where A, B and m are constants, while the former does not.

Progress in the field would probably be best aided by a joint effort--whether in the format of a workshop or a task-force study--of experimental investigators, mechanics experts, and practicing engineers to formulate the needs for various conditions and the corresponding necessary tests. Such an activity would not, of course, immediately provide all the desired answers, but would probably focus future research in the most useful directions.

Attention has been centered here on the importance of the subject from the standpoint of the calculation of the mechanical response; one should however not ignore the problems presented by the temperature-dependence of properties in the simple diffusion problem itself. The prevailing feeling probably is that (once the variation of the various properties with temperature has been determined) such problems can only be dealt with by means of numerical solutions. Even so, most numerical analyses ignore the temperature dependence because of the difficulties introduced by the additional non-linearity, and certainly more research towards rendering solutions of the type practicable would be useful.

It may be noted, in the meantime, that work is lacking even in the simple Neumann case, in which, as has been mentioned, similarity holds even if the properties are temperature-dependent. In that case [2.4] the construction of the solutions as linear combinations of elementary ones with stationary boundaries should prove useful for the numerical analysis of many simple but important change-of-phase problems.

4. APPROXIMATIONS

Two types of approximations should be distinguished: the first are in the nature of basic assumptions underlying the physical model adopted for the solution, the second refers to the method of solution itself. A few words about each of these types may be of interest.

Basic assumptions, introduced as a matter of course in order to bring complex actual problems within reach of solution, are normally justified by <u>ad hoc</u> considerations of the particular problem at hand. As a consequence, their validity cannot be judged except by someone with considerable acquaintance and expertise with the field. Furthermore, they are not always very explicitly stated in most papers, since these are as a rule addressed to other experts in the same field, namely to people whose familarity with them makes their repetition superfluous. Thus, for example, papers describing moving-boundary problems in astrophysics (e.g., [4.6,21]) or in combustion (e.g.,[4.16]) are difficult to assess by workers with primary interest, say, in metal castings. What is more serious is the fact that this makes effective communications between different groups difficult; such communication as does occur in fact takes place at the next level of the investigation, namely after the physical model formulation has been accepted. At this level, the second type of approximation earlier mentioned comes into play.

Before proceeding to a discussion of methods of solution, however, one additional observation might be appropriate. The physical model must be chosen so as to represent with reasonable accuracy the aspects of the actual behavior which are of particular interest, and may slight others of lesser immediate interest. Thus, for example, it is permissible to neglect the change in density during change of phase if the progress of the solid-liquid interface is of paramount interest, but this effect cannot be ignored in determinating the details of the flow in the liquid [4.13] or the grain structure of the solid metal [4.14]. One particular example can illustrate this point rather dramatically: the rate of melting of the interior of nuclear reactor fuel rod can be described rather accurately [4.17] without consideration of the density increase upon melting, but if this effect is ignored, the catastrophic collapse caused by the accompanying high internal pressure will not appear [4.19,20].

Once the basic model to be studied has been established, one is ready to consider the method of solution. It goes without saying that this should be as accurate as possible, so as to insure that a faithful representation of the model's behavior has been obtained. The few cases in which an exact solution is possible require no further comment; for the others, a choice of numerical or analytical techniques, or a combination of them, must be made. It may seem inappropriate, especially in view of the separation of Methods and Applications in the format of this Conference, that in a survey of the latter a discussion of the former should be included, but I did think it was useful to mention the perhaps obvious fact that the method of solution must be fitted (just as the physical model was) to the practical goals of the analysis.

Numerical methods for Stefan-type problems will be discussed in detail by several other participants during this Conference. From the standpoint of applications, there is little doubt that numerical methods are the most attractive, and often the only ones, which can accomodate in any reasonable manner the inevitable practical complexities of the problem. This implies that emphasis should be placed on the development of numerical methods capable of incorporating the additional non-linearity introduced by the temperature dependence of material properties. Cooperation between analytical and numerical solutions is however extremely important; an example of this is the necessity of developing in many cases analytical short-time solutions to be extended numerically [4.20].

Approaches based on various finite difference schemes have been generally more popular in heat and mass transfer problems than those based on the finite element approach. The usefulness of the latter should however be further examined, particularly in problems in which the mechanical response is also desired.

Not to be overlooked among numerical methods are those which are based on essentially analytical developments, such as the numerical solution of integral equations (e.g., [4.11,18]).

We finally turn to approximate methods proper, namely techniques for developing approximate analytical solutions. These should play an important role in applied problems, since they can provide, at their best, rapid and hopefully reasonable estimates of the principal characteristics of the sought-for solutions. Best known among such methods is the "heat balance" approach [4.10], normally used in conjunction with a "penetration-depth" concept, namely an unknown distance from the exposed surface beyond which the body is unaffected by the conditions to which the surface is exposed.

The heat balance method is a special case of the "weighting-function" approach [2.6]; alternatively, it can be thought as a special case of a variational approach. The latter have received considerable attention in the literature either in direct form (e.g., [4.5]) or in Biot's Lagrangian formulation [4.1].

These methods have been rather successful in the solution of several one-dimensional problems. From the applied standpoint, more work seems to be necessary in three principal directions, namely the development of techniques for obtaining improvements on a given approximate solution, the adaptation to cases of variable properties, and their use in multidimensional problems. Some work on the first two of these is being done (e.g., [4.2,4]); application to moving-boundary problems of the Stefan-type was carried out in [4.7] (see also [2.2]) based on the use [4.4] of multiple penetration depth, and shows the expected improvement. An approximate method of solution of multidimensional problems based on a separation of variables approach was given in [4.12].

The accuracy of approximate solutions obtained by methods such as the ones mentioned above is always difficult to assess. When two or more successive approximations are available, one may be satisfied (or may have to be) in an engineering sense with a comparison between them. In the case of temperature dependent properties, it was suggested in [4.4] that comparison of two separate solutions of the problem (one obtained directly from the heat conduction equation, and the other after application of the Kirchhoff transformation) would yield a means of estimating the accuracy of either solution.

Finally, methods of determining strict but easily calculated bounds have been explored as a possible answer to this problem (e.g.,[4.3,9,15],but clearly more work in this direction is necessary. Published work in this area has mainly been based on theorems comparing the behavior under different prescribed surface conditions, or different rates of ablation, (cf.[2.6]), or on some special characteristics of the solution (e.g.,[4.15]. The development of further such theorems and their applications should be further examined; at the same time comparison theorems based on different variations of material properties should be devised. Such theorems would permit the replacement of the actual temperature dependence of the properties by simpler ones for which the solution is more tractable, and would thus provide useful and hopefully easily obtainable bounds to the desired solution.

Unfortunately, theorems of this type are difficult to develop, and in fact, to the author's knowledge, now have appeared in the literature. To indicate what might be done, the following proof, although allowing only limited variations of properties, may be an interesting first step illustrating both the opportunities and the difficulties of the problem.

Consider the one-phase Stefan problem in a slab $0<x<L$, initially solid at zero temperature, heated by a non-decreasing heat flux $Q_o(t)$ at $x = 0$ and insulated at $x = L$. The problem is thus described by the system

$$[K(T)T']' = (\rho c)(T)\dot{T} \qquad s(t) < x < L \tag{4.1}$$

$$T(x,0) = 0 \qquad o < x < L \tag{4.2}$$

$$T'(L,t) = o \qquad o < t < t_L \tag{4.3}$$

$$-K(T_m)T'(s,t) = Q_o(t) - \rho l\dot{s} \qquad o < t < t_L \tag{4.4}$$

$$T[s(t),t] = T_m \qquad t_m < t < t_L \tag{4.5}$$

$$s(t) = o \qquad 0 < t \le t_m \tag{4.6}$$

where the times t_m and t_L are defined by

$$T(0,t_m) = T_m \quad \text{and} \quad s(t_L) = L \tag{4.7}$$

The solution of this problem is unique [4.12, 2.6]; furthermore $\dot{T} \ge 0$ and $\dot{s} \ge 0$ throughout.

Consider now two problems, denoted by the subscripts 1 and 2 respectively, in which

$$(\rho c)_2(T) \ge (\rho c)_1(T) \tag{4.8}$$

Then, from the difference in the two heat balances

$$\int_o^t Q_o(t)dt = \int_{s_{1,2}(t)}^L H_{1,2}dx + (\rho l + H_{m_{1,2}})s_{1,2} \tag{4.9}$$

where the heat content is

$$H(T) = \int_o^T \rho c dT \quad ; \quad H_m = H(T_m) \tag{4.9a}$$

we have (cf.[4.12])

$$t_{L1} \le t_{L2} \tag{4.9b}$$

Assume now that the difference in the physical properties characterizing the two problems is such that the solutions of problem (4.1-7) satisfies

$$T_1 \ge T_2 \tag{4.10}$$

if the boundary conditions and the initial conditions satisfy inequalities of the type

$$T_1(s^*,t) \ge T_2(s^*,t) \quad \text{or} \quad T_1'(s^*,t) \le T_2'(s^*,t)$$
$$T_1(o,t) \ge T_2(o,t) \tag{4.11}$$

for any curve $0 < x = s^*(t) < L$. The development of useful and fairly general conditions for (4.10) to hold is however one of the principal difficulties of the problem, and no attempt to resolve it here is presented; we shall devote our attention to proving that, if it holds, then

$$s_1(t) \geq s_2(t) \tag{4.12}$$

Some simple conditions under which (4.10) is valid may be listed first. For example, if K is constant and equal in the two solutions, then the heat equation is

$$KT''_{1,2} = (\rho c)_{1,2}(T_{1,2})\dot{T}_{1,2} \tag{4.13}$$

and (4.10) follows from subtraction of one of (4.13) from the other, i.e.,

$$\begin{aligned} &K(T_2 - T_1)'' - (\rho c)_2(T_2 - T_1)^{\cdot} + \\ &[\rho c_1(T_1) - \rho c_2(T_2)]\, \dot{T}_1 = 0 \end{aligned} \tag{4.14}$$

provided that one can insure that

$$\rho c_2(T_2) \geq \rho c_1(T_1) \tag{4.14a}$$

This can be done if the specific heat is constant (or dependent on time and position but not temperature), or if c_2 is replaced by the maximum possible value of c_1 (which may be possible in change-of-phase problems because the extremes of temperature can often be estimated a priori).

As another example, let now

$$K_1(T) \geq K_2(T) \quad \text{and} \quad \kappa_1(T) \geq \kappa_2(T) \tag{4.15}$$

Then, with

$$T^*(T) = \frac{1}{K_o}\int_0^T K(T)dT \tag{4.16}$$

we have

$$\kappa(T^*)T^{*''} = \dot{T}^* \tag{4.16a}$$

and so

$$\varkappa_2(T_2^* - T_1^*)'' - (T_2 - T_1)^{\bullet} +$$

$$[\varkappa_2(T_2^*) - \varkappa_1(T_1^*)]T_1^{*\prime\prime} = 0 \qquad (4.17)$$

whence (4.10) again follows, if $T^{*\prime\prime} \geqq 0$, under conditions similar to those earlier considered, both for the effective temperatures T^* and the actual temperatures T.

As has been mentioned, we shall assume (4.10) to hold; in particular, it does so before change of phase starts, or

$$t_{m1} \leq t_{m2} \qquad (4.18)$$

In other words, (4.12) holds at $t = t_{m1}$ and at $t = t_{L1}$. It is obviously possible to devise functions $(\rho c)_2^*(T)$ or $K_1^*(T)$ sufficiently large, in the previous two examples respectively, in which (4.12) holds for all $t_{m1} < t < t_{L1}$

If (4.12) is to fail, there must exist a period where $s_1 < s_2$, and therefore a first time at which $s_1 = s_2$. Consider now a continuous sequence of functions bounded by $(\rho c)_2$ and $(\rho c)_2^*$, or by K_1 and K_1^*; There must exist one such function for which the period just mentioned shrinks to a point $t = t^*$ at which

$$s_1 = s_2 \quad \text{and} \quad \dot{s}_1 = \dot{s}_2 \qquad (4.16)$$

At that point (in either example, with T^* replacing T in the second),

$$T_1' = T_2' \;;\; \dot{T}_1 = \dot{T}_2 \;;\; T_1'' < T_2'' \qquad (4.16a)$$

respectively from (4.4), from the differential form $T'\dot{s} + T = 0$ of (4.5), and from either (4.13) or (4.16a). But then, near $x = s$ we have

$$T_{1,2} = T_m + T_{1,2}'(x - s) + T_{1,2}''[(x - s)^2/2] + \ldots \qquad (4.17)$$

or

$$T_1 - T_2 = (T_1'' - T_2'')[(x - s)^2/2] + \ldots < 0 \qquad (4.17a)$$

which contradicts (4.10); hence (4.12) must hold at all times.

5. MULTIDIMENSIONAL PROBLEMS

The major portion of the published work on moving boundaries is concerned with one-dimensional solutions, and indeed the mathematical complexities of the subject are such that many unsolved problems remain even with this restriction. Nevertheless, many practical problems are not, and cannot be, reasonably approximated by one-dimensional models. Many numerical methods are readily applied to multidimensional problems, but there appears to be considerable scope for analytical work, both approximate and exact, in this area.

It may be useful to recall the classification of multidimensional problems presented in [5.13], in which a distinction was made according to the type of motion characterizing the spread of the change-of-phase interface along the surface. Class I includes problems in which the change of phase starts simultaneously at all points of the surface. One-dimensional problems are of this class, as are all solutions in which the initial conditions are ignored or in which a steady-state is assumed to prevail. This class includes also the important two-dimensional solutions of [5.8,9] pertaining to dendritic growth and the work of [5.11] on the effect of imperfect mold-ingot contact.

Class II comprises problems in which change of phase starts simultaneously at all points in some portion of, but not the entire surface, and Class III those in which it starts at a point. In these two classes the intercept between the phase interface and the surface moves along the surface, at the same time as the interface itself progresses into the medium; as a consequence these problems are of a higher order of magnitude of difficulty than those of Class I (cf. [5.1,3,10] and the references cited there). Nevertheless, problems of Class III are some of the physically most important ones, particularly at the early stages of the various processes being examined under the general heading of moving boundaries , and should be attracting more attention on the part of researchers. Particularly lacking are works in which more than one phase is considered, and those with more complicated interface conditions. Among the latter, a particularly interesting effect often neglected is that of curvature on the interface equilibrium temperature [5.5,12] and on concentration [5.10], usually expressed by the Gibbs-Thompson relations

$$T_r = T_\infty - \frac{2\sigma T_m}{\ell r} \quad ; \quad C_r = C_\infty\left(1 + \frac{2\sigma V}{RTr}\right) \tag{5.1}$$

or its generalizations [5.4,6,10]. In these equations the subscripts r and ∞ indicate equilibrium values corresponding to surfaces with radius of curvature r and plane respectively, σ is the surface energy, V the molar volume of the solute or crystal, and R the gas constant. If the coupled heat and mass transfer problem is analyzed, the effects on both temperature and concentration must of course be included simultaneously.

The extension of the standard analytical approximate techniques--heat balance, Biot's method and the like--into multidimensional problems has hardly been explored, and perhaps not found feasible. The same can be said of the construction of practical ways of devising bounds for the solution of such problems. Some promising approximate methods for multidimensional problems were presented in [4.7], based either on the concept of a thermal layer or on simplification in the heat equation.

Development of bounds on the basis of theorems comparing the behavior of one or two phase problems of the type considered for the one-dimensional case in [5.2,14,15] would be useful. This would particularly be true if the analytical solution were to be used as a starting solution for further numerical work (i.e., for very short times), since it may be conjectured that the one-dimensional conclusion that the starting solution is the same in these problems will carry over to multidimensional problems. If that is true, then one may obtain a valid starting solution for all cases from the obviously simpler one-phase case.

6. MACROSCOPIC AND MICROSCOPIC APPROACHES

The preceding discussion has been exclusively concerned with solutions based on the continuum approach; at the same time considerable research is being pursued on such topics as the structure of growing crystals, which clearly require analyses at the atomic level of the material (e.g.,[6.4-6,12]. Consideration of both approaches and of their relation is evident in some works (e.g.,[6.10]), and leads to a formulation more complete than has formally been considered in most available analyses. Such a formulation requires the temperatures T_S and T_L, and the concentrations C_S and C_L, in the solid and liquid respectively, to satisfy the heat and diffusion equations under interface conditions of the following type during solidification from the melt:

$$C_S = kC_L \; ; \; D_S \frac{\partial C_S}{\partial n} - D_L \frac{\partial C_L}{\partial n} = V_n (1 - k) C_L \tag{6.1}$$

$$T_S = T_L \; ; \; K_S \frac{\partial T_S}{\partial n} - K_L \frac{\partial T_L}{\partial n} = V_n \rho \ell \tag{6.2}$$

where V_n is the normal component of the interface velocity and k the partition coefficient. The usual way of determining V_n[6.6] simplifies the appropriate expression by means of fact that the temperatures at the interface are constant and equal to the melting temperature. It has already been seen, however, that this condition is not strictly correct for curved interfaces: the equilibrium temperature there must satisfy a thermodynamic relation of the form

$$T_r = T_E(C_L) - \frac{\sigma}{Sr} \tag{6.3}$$

where S is the entropy of melting and where T_E is the equilibrium freezing temperature of a plane surface at concentration C_L, the functional relation between T_E and C_L being determined by the liquidus line of the phase diagram for the material in question. The additional relation required to complete the formulation is then the so-called kinetic condition, which relates the velocity V of freezing to the difference between the interface and the equilibrium temperatures, or

$$V = f(T - T_r) \tag{6.4}$$

The form of this relation depends on the details of the atomic mechanism of crystal formation [5.5] and is usually taken to be linear (e.g.,[5.12], [6.10], or

$$V = \mu(T - T_r) \tag{6.5}$$

where μ is the atomic kinetic coefficient.

Solution of the complete problem outlined above has not yet been attempted, and it should in fact be noted that it is not expected to be unique on physical grounds. In other words, several interface shapes are thought to exist which satisfy the formulation given (the initial and fixed-boundary conditions must of course be added); the actual one is the one which corresponds to the maximum velocity of growth, or [6.10]

$$\delta V_{ext} < 0 \tag{6.6}$$

where V_{ext} is the velocity of the extremities of the formed crystal, and δ indicates the difference in the velocities corresponding to the actual and any other interface morphologies.

It would seem desirable to perform some careful studies of uniqueness of solution of the stated problem; in particular, such studies might shed some light on the types of functional relations (6.4) which are theoretically admissible, thus simplifying (or at least providing some direction for) the task of the experimentalists. Of course, if the simple one-dimensional problem is considered, the curvature dependence and thus most of the foregoing difficulties are absent; uniqueness in this case has been considered [6.2]. In that work, the coupled heat and mass transfer problem is considered; similarity solutions for this problem have been examined by a number of authors (e.g., 6.8-10]), and a more general case has been discussed in the literature [6.1].

One additional difficulty should be added to the foregoing formulation, namely that required to describe the generally anisotropic character of the formed crystals [5.5, 6.7]. This effect has not been considered before, but it can be safely expected to complicate matters a great deal, because the orientation of the crystals is not known a priori, but must be determined by a condition similar to (6.6).

It might be worth while to consider here the modification which anisotropy would introduce in the known exact short-time solution for the half-space $z > 0$ under arbitrary heating rates [5.5]; this is easy to do if one assumes z to be a principal direction of the conductivity tensor. Consider for simplicity the two-dimensional special case of the problem [6.3] and let the x and z directions be principal with conductivities K_1 and K_3 respectively. It is then easy to show that the transformation

$$x' = x\sqrt{K_3/K_1} \quad ; \quad z' = z \quad ; \quad s' = s \tag{6.7}$$

reduces the entire problem to its original form, since the interface heat-balance condition

$$-\left[K_3 + K_1\left(\frac{\partial s}{\partial x}\right)^2\right]\frac{\partial T}{\partial x} = Q_o(x,t) - \rho\ell\frac{\partial s}{\partial t} \tag{6.8a}$$

becomes

$$-K_3\left[1 + \left(\frac{\partial s'}{\partial x'}\right)^2\right]\frac{\partial T}{\partial z'} = Q_o'(x',t) - \rho\ell\frac{\partial s'}{\partial t} \tag{6.8b}$$

Hence the solution is obtained simply by replacing the quantity $C_{2x} = (\partial^2T/\partial x^2)$ by $C_{2x'} = -(\partial^2T/\partial x'^2)$, evaluated at the instant of start of melting. Thus

$$C_{2x'} = C_{2x}(K_1/K_3) \tag{6.9}$$

Thus the maximum penetration is

$$\frac{s_{max}(t)}{2\sqrt{\kappa_3 t_m}} = \frac{4MC_1}{3\pi}\left[1 + \frac{1}{8}\left(\frac{C_{2x}K_1}{C_1K_3}\right) + \ldots\right]T^{3/2} + \ldots \tag{6.10a}$$

and the maximum surface extent of melting is

$$\frac{x_o(\tau)}{2\sqrt{\kappa_3 t_m}} = \sqrt{\frac{2C_1\tau}{C_{2x}}}\left[1 - \frac{1}{8}\left(\frac{C_{2x}K_1}{C_1K_3}\right) + \frac{11}{128}\left(\frac{C_{2x}}{C_1}\frac{K_1}{K_3}\right)^2 + \ldots\right] + \ldots \quad (6.11)$$

where

$$M = \frac{\sqrt{\pi c}\ T_m}{2\ell}\ ;\quad \kappa_3 = \frac{K_3}{\rho c}\ ;\quad \tau = \frac{t - t_m}{t_m}\ ;\quad C_1 = \left(\frac{\partial T}{\partial t}\right)_{\substack{t=t_m \\ x=z=0}} \quad (6.12)$$

It is of interest to calculate the ratio of the two distances given in eqs. (6.10), or

$$\frac{s_{max}(\tau)}{x_o(\tau)} = \frac{4M\sqrt{C_1C_{2x}}}{3\pi\sqrt{2}}\ \tau\left[1 + \frac{1}{4}\left(\frac{C_{2x}}{C_1}\frac{K_1}{K_3}\right) + \ldots\right] + \ldots \quad (6.13)$$

This appears at first sight to predict that the penetration will be greater (relative to the surface spread) the greater the ratio of conductivities (K_1/K_3) in the directions x_o and s respectively. This would be intuitively disturbing; in reality, a true assessment of the effect of variations in conductivity, is obtained only if the accompanying variation in the pre-melting solution is included. A meaningful comparison on the basis of eq. (6.13) therefore results only if it is assumed that the temperature distribution at the instant of incipient melting is unchanged. Thus, consider two cases corresponding to conductivities K_1' and K_1'' (with K_3 unaltered for conceptual simplicity); then it is necessary that

$$K_1'C_{2x}' = K_1''C_{2x}'' \quad (6.13)$$

or that C_{2x} be inversely proportional to K_1. Substitution in (6.13) will then give the more acceptable result that s_{max}/x_o decreases with increasing values of (K_1/K_3).

7. CONCLUDING REMARKS

The foregoing discussion has attempted to indicate--although in an admittedly sketchy and far from comprehensive manner--a number of problems connected with moving boundaries which have not received great emphasis during this conference. In the presentation an effort has been made to identify areas in which further cooperation between mathematicians and engineers is most desirable and can most readily be expected to bear fruit. It is hoped that this brief exposition will be helpful in furthering that cooperation.

REFERENCES

1.1 J.R. Ockendon and W.R. Hodgkins, Editors: Moving Boundary Problems in Heat Flow and Diffusion, Clarendon Press, Oxford, 1975, being the Proceedings of a Conference held at the University of Oxford (March 1974).

2.1 B.A. Boley, "A General Starting Solution for Melting and Solidifying Slabs," Int. J. Eng. Sci., vol. 6, 89-111 (1968).

2.2 B.A. Boley, "Applied Thermoelasticity," in Developments in Mechanics, vol. 7, Proceedings of 13th Midwestern Mechanics Conference, University of Pittsburgh. Editors: J.L. Abrams and T.C. Woo, 1-16 (1973).

2.3 B.A. Boley, "Bounds on the Maximum Thermoelastic Stress and Deflection in a Beam or Plate," J. Appl. Mech., vol. 33, no. 4, 881-887 (December 1966).

2.4 B.A. Boley, "On a Melting Problem with Temperature-Dependent Properties," in Trends in Elasticity and Thermoelasticity, Witold Nowacki Anniversary Volume, Wolters-Nordhoff Publ. Co., The Netherlands, 21-28 (1971).

2.5 B.A. Boley, "Temperature and Deformations in Rods and Plates Melting under Internal Heat Generation," Proc. First Int. Conf. on Struct. Mech. in Reactor Technology (SMiRT-1), Berlin, vol. 6, Section L, Paper L2/3 (1972).

2.6 B.A. Boley, "The Analysis of Problems of Heat Conduction and Melting," in High Temperature Structures and Materials, edited by A.M. Freudenthal, B.A. Boley & H. Liebowitz, Pergamon Press, 260-315 (1964).

2.7 B.A. Boley and J.H. Weiner, Theory of Thermal Stresses, J. Wiley & Sons, N.Y. (1960).

2.8 H.S. Carlsaw and J.C. Jaeger, Conduction of Heat in Solids, Second Ed., Oxford University Press (1959).

2.9 E. Friedman and B.A. Boley, "Stresses and Deformations in Melting Plates," J. Spacecraft, vol. 7, no. 3, 324-333 (March 1970).

2.10 Y. Masuko and S. Matsunaga, "On the Temperature Distribution in Solidifying Ingots and the Residual Stresses," Sumitomo Kinzoku Journal, vol. 18, no. 1 41-51 (Jan.1966).

2.11 O. Richmond and N.C. Huang, "Interface Instability during Unidirectional Solidification of a Pure Metal," Proc. Sixth Canadian Congress of Applied Mechanics (CANCAM) Vancouver, May 29-June 3 (1977).

2.12 O. Richmond and R.H. Tien, "Theory of Thermal Stresses and Air-Gap Formation during the Early Stages of Solidification in a Rectangular Mold," J. Mech. Phys. Solids, vol. 19, 273-284 (1971).

2.13 O.T.G. Rogers and E.H. Lee, "Thermo-Viscoelastic Stresses in a Sphere with an Ablating Cavity," Brown University, Tech. Rep., No. 6, Contract NOrd - 18594 (August 1962).

2.14 J. Savage, "A Theory of Heat Transfer and Air Gap Formation," J. Iron & Steel Institute, vol. 200, no. 1, 41-47 (January 1962).

2.15 F.L. Schuyler and E. Friedman, "High Temperature Ablation Interaction," article in Mechanics of Composite Materials, Editors F.W. Wendt, H. Liebowitz, and N. Perrone, Pergamon Press, New York, 769-798(1970).

2.16 I. Tadjbakhsh, "Thermal Stresses in an Elastic Half-Space with a Moving Boundary," AIAA Journal, vol. 1, no. 1, 214-215 (January 1963).

2.17 R.H. Tien and V. Koump, "Thermal Stresses during Solidification on Basis of an Elastic Model," J. Appl. Mech., vol. 36, 763-767 (1969).

2.18 J.H. Weiner and B.A. Boley, "Elasto-Plastic Thermal Stresses in a Solidifying Body," J. Mech, and Phys. of Solids, vol. 11, 145-154 (1963).

2.19 P.J. Wray, "Non-Uniform Growth of a Plate on a Chilled Surface," AIME Annual Meeting, Atlanta, March 6-10 (1977).

3.1 Y.S. Touloukian and C.Y. Ho, Editors: Thermophysical Properties of Matter, TPRC Data Series, Plenum Publ. Corp., New York.

3.2 P.J. Wray, "Plastic Deformations of Delta-Ferritic Iron at Intermediate Strain Rates," Metell. Trans A, vol. 7A, 1621-1627 (November 1976).

3.3 P.J. Wray and M.F. Holmes, "Plastic Deformations of Austenitic Iron at Intermediate Strain Rates," Metell. Trans, A, vol. 6A, 1189-1196 (June 1975).

4.1 M.A. Biot and H. Daughaday, "Variational Analysis of Ablation," J. Aerospace Sci., vol 29, 227-229(1962).

4.2 B.A. Boley, "On the Use of Superposition in the Approximate Solution of Heat Conduction Problems,"Int. J. Heat & Mass Transfer, vol. 16, 2035-2041 (1973).

4.3 B.A. Boley, "Upper and Lower Bounds for the Solution of a Melting Problem," Quart, Appl. Math., vol. 21, no. 1, 1-11 (1963).

4.4 B.A. Boley and L. Estenssoro, "Improvements on Approximate Solutions in Heat Conduction," Mechanics Research Communications, vol. 4, no. 4, 271-279 (1977).

4.5 G. Duvaut, "The Solution of a Two-Phase Stefan Problem by a Variational Inequality," article in [1.1], 173-181 (1974).

4.6 P.P. Eggleton, "A Moving Boundary Problem in the Study of Stellar Interiors," article in [1.1], 103-111(1974).

4.7 L.F. Estenssoro, "Some Approximate Solutions of Heat Conduction Problems," M.S. Thesis, Northwestern University, August 1976.

4.8 L. Fox, "What are the Best Numerical Methods?," article in [1.1], 210-241 (1974).

4.9 D. Glasser and J. Kern, "Bounds and Approximate Solutions to Linear Problems with Nonlinear Boundary Conditions: I - Solidification of a Slab," AIChE Journal, in press (1977).

4.10 T.R. Goodman, "Application of Integral Methods to Transient Nonlinear Heat Transfer," Advances in Heat Transfer, vol. I, Academic Press, 51-122 (1964).

4.11 P. Grimado and B.A. Boley, "A Numerical Solution for the Axisymmetric Melting of Spheres," J. Numerical Methods in Eng., vol. 2, 175-188 (1970).

4.12 Y. Horie and S. Chehl, "An Approximate Method of Solution for Multidimensional Crystal Growth Problems," J. Crystal Growth, vol. 29, 248-256 (1975).

4.13 G. Horvay, "Freezing into an Undercooled Melt Accompanied by Density Change," Proc. Fourth Nat. Congress of Applied Mechanics, ASME, New York, 1315-1325 (1962).

4.14 G. Horvay, "The Tension Field Created by a Spherical Nucleus Freezing into Its Less Dense Undercooled Melt," Int. J. Heat & Mass Transfer, vol. 8, 195-243 (1965).

4.15 J. Kern, "A Simple and Apparently Safe Solution to the Generalized Stefan Problem," Int. J. Heat & Mass Transfer, vol. 20, no. 5, 467-474 (May 1977).

4.16 H. Krier, J.S. Tien, W.A. Sirignano, and M. Summerfield, "Nonsteady Burning Phenomena of Solid Propellants: Theory and Experiment," AIAA Journal, vol. 6, no. 2, 278-285 (February 1968).

4.17 A. Lahoud and B.A. Boley, "Some Considerations on the Melting of Reactor Fuel Plates and Rods," Nuclear Engineering and Design, vol. 32, 1-19 (1975).

4.18 J.M. Lederman and B.A. Boley, "Asixymmetric Melting or Solidification of Circular Cylinders," Int. J. of Heat & Mass Transfer, vol. 13, 413-427 (1970).

4.19 M.F. Lyons, R.C. Nelson, T.J. Pashos, and B. Weidenbaum, "U_2O Fuel Rod Operation with Gross Central Melting," Trans, Am. Nucl. Soc., vol. 6, 155-156 (1963).

4.20 T.J. Pashos and M.F. Lyons, "U_2O Fuel Rod Performance with a Molten Central Core," Trans. Am. Nucl. Soc., vol.5(2), 462 (1962).

4.21 D.L. Turcotte and G. Schubert, "Structure of the Olivine-Spinel Phase Boundary in the Descending Lithosphere," J. Geophysical Research, vol. 76, no. 32, 7980-7987 (November 1971).

5.1 B.A. Boley, "The Embedding Technique in Melting and Solidification Problems," article in [1.1], 150-172 (1974).

5.2 B.A. Boley, "Upper and Lower Bounds in Problems of Melting and Solidifying Slabs," Quart. J. Mech & Appl. Math., vol. 17, Part 3,253-269 (1964).

5.3 B.A. Boley and H.P. Yagoda, "The Three-Dimensional Starting Solution for a Melting Slab," Proc. Roy. Soc. London, Series A, 89-110 (1971).

5.4 N. Cabrera and R.V. Coleman, "Theory of Crystal Growth from the Vapor," article in [6.5], 1 - 28 (1963).

5.5 B. Chalmers, Principles of Solidification, J. Wiley & Sons, New York (1964).

5.6 C. Herring, "Structure and Properties of Solid Surfaces," Eds. R. Gomer and C.S. Smith, Univ. of Chicago, p.5 (1953).

5.7 C. Cheering, Symposium on the Physics of Powder Metallurgy, McGraw-Hill Co., New York (1949).

5.8 G. Horvay and J.W. Cahn, "Dendritic and Spheroidal Growth," Acto Metall., vol. 9, 695-705 (July 1961).

5.9 G.P. Ivantsov, Dokl. Akad. Nauk, SSSR, vol 58, 567 (1947).

5.10 J.W. Martin and R.D. Doherty, Stability of Microstructure in Metallic Systems, Cambridge University Press, Cambridge (1976).

5.11 P.D. Patel and B.A. Boley, "Solidification Problems with Space and Time Varying Boundary Conditions and Imperfect Mold Contact," Int. J. Eng. Sci., vol. 7, 1041-1066 (1969).

5.12 R.J. Schaefer and M.E. Glicksman, "Fully Time-Dependent Theory for the Growth of Spherical Crystal Nuclei," J. of Crystal Growth, vol. 5, 44-58 (1969).

5.13 D.L. Sikarskie and B.A. Boley, "The Solution of a Class of Two-Dimensional Melting and Solidification Problems," Int. J. Solids and Structures, vol. 1, no. 2, 207-234 (1965).

5.14 T.S. Wu, "Bounds in a Melting Slab with Several Transformation Temperatures," Quart. J. Mech. & Appl. Math., vol 19, part 2, 183-195 (1966).

5.15 T.W. Wu and B.A. Boley, "Bounds in Melting Problems with Arbitrary Rates of Ablation," SIAM J. Appl. Math., vol. 14, no. 2, 306-323 (1966).

5.16 H.P. Yagoda and B.A. Boley, "Starting Solutions for Melting of a Slab under Plane or Axisymmetric Hot Spots," Quart. J. Mech. Appl. Math., vol. 23, part 2, 225-246 (1970).

6.1 B.A. Boley, "Time-Dependent Solidification of Binary Mixtures," to appear in Int. J. Heat & Mass Transfer (1978).

6.2 B.A. Boley, "Uniqueness of Solution of a Coupled Heat and Mass Change of Phase Problem," in preparation.

6.3 B.A. Boley and H.P. Yagoda, " The Starting Solution for Two-Dimensional Heat Conduction Problems with Change of Phase," Quart. Appl. Math., vol. 27, no. 2, 223-246 (1969).

6.4 G.A. Chaduick, Metallography of Phase Transformations, Crane, Russak & Co., New York (1972).

6.5 J.J. Gilman, Editor, The Art and Science of Growing Crystals, J. Wiley and Sons, New York (1963).

6.6 R.A. Laudise, The Growth of Single Crystals, Prentice-Hall, Englewood Cliffs, N.J. (1970).

6.7 G.E. Nash, "Capillarity-Limited Steady-State Dendritic Growth; Part 1 - Theoretical Development," NRL Report 7679, Naval Research Laboratory, May 24, 1974, p.17.

6.8 P.D. Patel, "Interface Conditions in Heat Conduction Problems with Change of Phase," AIAA Journal, vol. 6,2454(1968).

6.9 A.B. Taylor, "The Mathematical Formulation of Stefan Problems," article in [1.1], 120-137 (1974).

6.10 W.A. Tiller, Principles of Solidification, article in [6.5], pp.276-312 (1963).

6.11 T. Tsubaki and B.A. Boley, "One-Dimensional Solidification of Binary Mixtures," Mechanics Research Communication, vol. 4, no. 2, 115-122 (1977).

6.12 John D. Verhoeven, Fundamentals of Physical Metallurgy, J. Wiley & Sons, New York (1975).

6.13 C. Wagner, "Theoretical Analysis of Diffusion of Solutes during the Solidification of Alloys," Journal of Metals, Trans, AIME, vol. 200, 154-160 (February 1954).

Bruno A. Boley
Dean, Technological Institute
Northwestern University
2145 Sheridan Road
Evanston, Illinois 60201

This work was performed with support from the Office of Naval Research

MOVING BOUNDARY PROBLEMS

MODELING OF MOVING BOUNDARIES DURING SEMICONDUCTOR FABRICATION PROCESSES

R. W. Dutton, D. A. Antoniadis

The moving boundary problem for oxide growth in silicon surfaces is described. The one-dimensional problem involves a known oxide interface velocity which is controlled by ambient pressure and temperature. Interface conditions for dopant impurity distributions both in the oxide and in silicon are modeled and simulated. An algorithm based on the impurity conservation equation is described. Detailed considerations for one-dimensional grid allocation and the contributions of moving boundary induced impurity flux are discussed. Recent experimental data for impurity segregation are discussed in terms of hypothetical physical models.

1. INTRODUCTION

The fabrication of Integrated Circuits (IC's) in silicon involves a sequence of steps which selectively change the electrical properties of the material. For example the thermal growth of oxide on the surface is essential both for electrical passivation and to serve as a mask for selective diffusion of impurities into the silicon through mask windows. Donor and acceptor atoms must be diffused into the silicon to achieve regions which conduct due to either excess electrons (n-type) or holes (p-type). For many high performance technologies uniformly doped epitaxial regions are chemically deposited over other selectively diffused regions.

Figure 1 summarizes in flow chart fashion a typical sequence of steps used to fabricate an integrated circuit for a bipolar transistor technology. Several points should be emphasized. First, at each Mask step a new layer of oxide must be grown during diffusion with sufficient thickness to guarantee that impurities do not penetrate regions without windows. Since each of these oxidation steps selectively consumes a fraction of the silicon, one can observe that the surface becomes terraced. This second point implies that oxidation for IC fabrication frequently involves two dimensional effects. Third, both

ISBN 0-12-757350-X

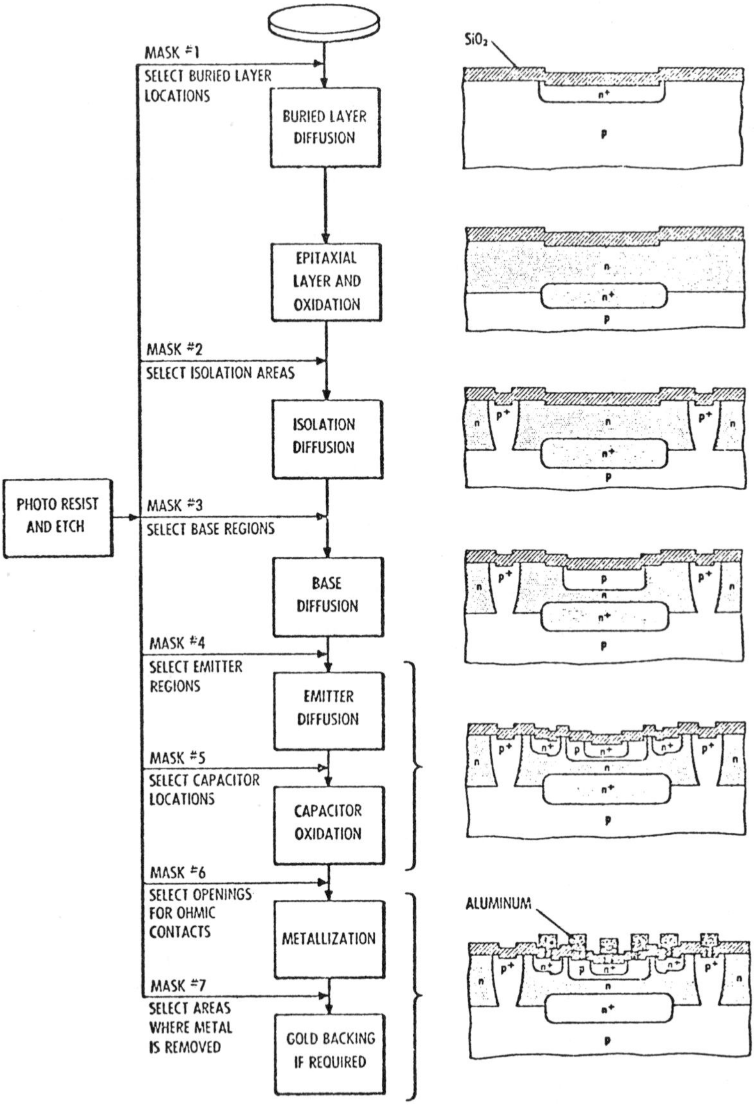

Figure 1. Flow chart of the sequence of processes used in the fabrication of a single-crystal monolithic circuit.

oxidation and epitaxial growth of silicon layers involve moving boundary effects whereby the impurity concentrations in both the existing silicon and growing layer are not constant. In addition, the surface and growth conditions can substantially alter these impurity distributions.

In this paper the details involved in the numerical simulation of impurity redistribution during silicon oxidation will be described. The mathematics and simplifying assumptions will be discussed. Finally, recent experimental work will be presented and discussed in the context of computer simulated results.

2. THE MOVING BOUNDARY OXIDATION PROBLEM

Since silicon-silicon oxide systems are the backbone of current semiconductor processing technologies, a significant amount of effort has been directed towards finding practical means of accurate impurity profile prediction in such systems. Several workers [1-3] have presented analytic solutions for the case of oxidation under special conditions. Such solutions do not generalize to arbitrary initial impurity distributions, non-uniform diffusivity, non-constant segregation coefficient, and non-parabolic oxide growth.

Early numerical solutions of the oxidation problem were based on a coordinate transformation [4] and/or numerical calculation of Green's functions [5], thus eliminating reference to the oxide. Although these techniques overcome many of the problems cited above, any effects related to the interaction between the oxide and the silicon are lost.

Accurate numerical solutions of the impurity redistribution under oxidation conditions, either for general purpose simulators [6] or for research tools in the investigation of the silicon-silicon oxide system require that both media be treated. Although a number of works in which such solutions are employed have been reported [7-9], only one [8] deals with specific problems associated with developing numerical algorithms. The equation of impurity conservation is used in integral form rather than in the usual differential form in the most recent efforts [8,9]. In the present work the implications of deriving numerical schemes from the integral conservation equation are studied.

3. THE CONSERVATION EQUATION

The general equation describing the conservation of particles in the absence of chemical generation or loss can be written as

$$\oint_{S(t)} \vec{F}.\vec{n}\ da = -\frac{d}{dt}\int_{V(t)} C\ dv \tag{1}$$

where

$\vec{F}$ = flux vector

$\vec{n}$ = outward unit normal

$S(t)$ = closed surface (function of time)

$V(t)$ = volume enclosed by $S(t)$

C = concentration

This equation states that the net number of particles leaving through the surface S is equal to the net time rate of change of the number of particles in the enclosed volume V. In general, the flux F is related functionally to the concentration C.

The formulation of numerical methods for the solution of (1) requires discretizations both in time and in space.

4. DISCRETIZATION IN TIME

The general form of (1) in time is given by

$$H(t) = \frac{d}{dt} G(t) \tag{2}$$

where

$$H(t) = \oint_{S(t)} \vec{F}.\vec{n}\ da$$

$$G(t) = -\int_{V(t)} C\ dv$$

It is well known that during oxidation of silicon there exists an impurity flux across the silicon-silicon oxide interface due to the motion of this interface [4]. This motion induced flux is given by

$$F_B(t) = -(1/m - \alpha) V_{ox} C_{si}(B,t) \tag{3}$$

where m is the thermodynamic segregation coefficient -- the ratio of impurity concentration in the silicon to that in the oxide at the interface -- and α is the ratio of oxidized silicon to resultant oxide volume. V_{ox} is the velocity of the moving interface relative to the oxide. It should be noted that an inherent assumption in (3) is that the segregation process is always in equilibrium. A more general form allowing the inclusion of first order kinetics is discussed in Section 8.

It is evident from (3) that at one point in space there exists a jump discontinuity of the impurity flux. This discontinuity moves through space during the oxidation. It is therefore apparent that the function H encounters a discontinuity when the oxide/silicon interface crosses the surface S. Consequently, straight forward approximations to (2) in differential form will not be correct since they do not include any reference to the possible discontinuity.

In order to avoid this problem and to include reference to the jump continuity methods based upon numerical integration [10] are chosen as the basis for numerical schemes to solve this problem. In general form, (2) is expressed as

$$\int_{t_o}^{t_1} H(t)\,dt = G(t_1) - G(t_o) \tag{4}$$

The first order implicit method then becomes for continuous H

$$H(t_1) = (G(t_1) - G(t_o))/(t_1 - t_o) \tag{5}$$

However, when a discontinuity occurs at a time t in the interval (t_o, t_1) as shown in Figure 2, the approximation to the integral yields

$$H(t_1) + \frac{(t_1 - t)}{(t_1 - t_o)} \Delta H = (G(t_1) - G(t_o))/(t_1 - t_o) \qquad (6)$$

where ΔH is given by equation (3)

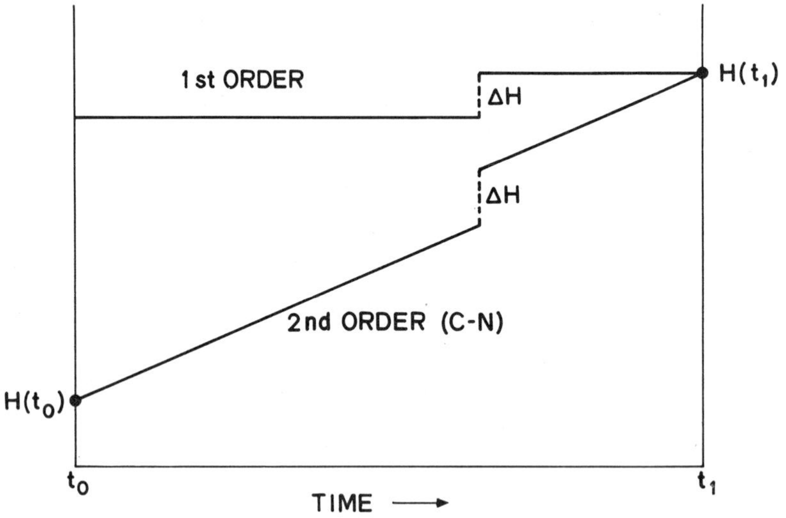

Figure 2. An example of approximating an integrand with a step discontinuity. An implicit first order approximation and a second order (Crank-Nicholson) approximation are shown.

If the implicit 2nd order method (Crank-Nicholson) is used to approximate (4) the result for continuous H is

$$\tfrac{1}{2} (H(t_1) + H(t_o)) = (G(t_1) - G(t_o)) / (t_1 - t_o) \qquad (7)$$

and for a single jump discontinuity in the interval (t_o, t_1) it is

$$\tfrac{1}{2} (H(t_1) + H(t_o)) + \frac{(t_1 - t)}{(t_1 - t_o)} \Delta H = (G(t_1) - G(t_o)/(t_1 - t_o) \qquad (8)$$

where again the jump occurs at t. The formulation of ΔH depends upon the spatial discretization but is derived from (3) for one dimension.

5. SPATIAL DISCRETIZATION

To perform the spatial discretization, the region over which (2) is to be solved is divided into subregions (subvolumes) and set of grid points (nodes) is defined. The functions H and G are then approximated in terms of node concentrations. The time variation of the subvolumes and subsurfaces must be taken into account when calculating the approximations and, as a result, time and space discretizations are intertwined.

The development of methods to approximate the spatial dependence of the solution must take into account the discontinuous behavior of the solution in space as well as changes in the physical structure near the Si/SiO_2 interface. In addition, it is necessary to consider the discrete subvolumes according to their relationship to the moving interface in order to determine when the motion induced flux, F_B, must be included. Since volumes and surfaces appear explicitly in the continuity equation (1) the methods developed here for one dimension can be adapted to the solution of two dimensional problems also.

6. THE METHOD IN ONE DIMENSION

Defining a one dimensional cartesian frame, the conservation equation (2) can be rewritten as

$$\int_{t_o}^{t_1} \left\{ F(x_2,t) - F(x_1,t) \right\} dt = - \left[\int_{x_1(t_1)}^{x_2(t_1)} C(x,t_1)\, dx - \int_{x_1(t_o)}^{x_2(t_o)} C(x,t_o)\, dx \right] \quad (9)$$

where x_1 and x_2 correspond to the surface S(t) in (1) and (2).

Figure 3 illustrates one method of discretization of the space occupied by the silicon and silicon oxide. Using this method, the moving SiO_2/Si interface always lies on a node and the nodes on either side remain a minimum specified distance from the interface node. The cell boundaries (x_1 and x_2 in (9) are then defined to be midway between the nodes. This technique avoids the problem of some cells becoming arbitrarily small compared to the majority.

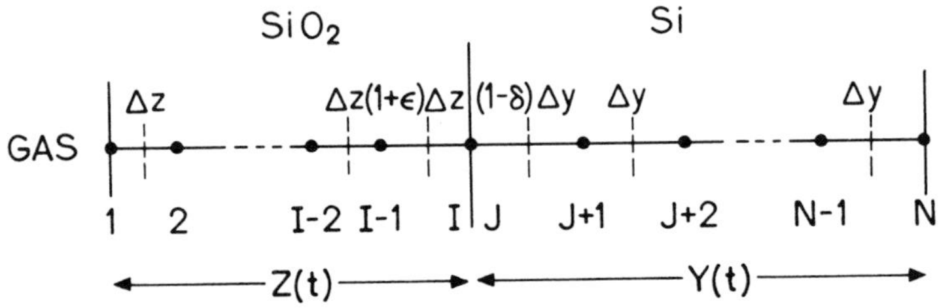

Figure 3. A method for discretization of the Si/SiO region. I and J are chosen so that $(I + \varepsilon)\ \Delta z = Z(t)$ and $-1/2 \leq \varepsilon < 1/2$. J and δ are chosen so that $(J + \delta)\ \Delta y = \delta Z(t)$ and $-1/2 \leq \delta < 1/2$. N remains fixed. The total number of nodes at any time is given by $N-J(t) + I(t) + 1$.

Once the discretization in space has been established it is necessary to combine it with the discretization in time. Given the constraint that the number of oxide nodes is allowed to increase by at most one node between time steps, four possibilities for advancement of the interface exist. Two correspond to the case when the number of oxide nodes remains constant as the interface advances while the other two correspond to the case when the number of oxide nodes increases by one. In both cases there exists the possibility that the interface crosses the left hand surface of the cell containing the interface at time t. Under these conditions the induced flux must be included in the discrete equations otherwise the induced flux should not appear since the moving interface did not coincide with any subsurface.

7. NUMERICAL RESULTS

In order to demonstrate the numerical methods presented here a benchmark example for which the analytic solution has recently been reported [3] is compared with numerical solutions obtained with and without the interfacial induced flux. Of particular interest is the behavior of the numerical solutions as a function of the impurity diffusivity ratio, D_{ox}/D_{si}, since many impurities have a very small diffusivity ratio. In addition, the behavior of the numerical solutions as a function of segregation coefficient, m, is important since it may vary significantly as a function of temperature for a given impurity.

The interface impurity concentration, C_{si}, and the amount of impurities leached from the silicon during oxidation, Q_{ox}, are shown for calculations made with and without the induced interfacial flux. The theoretical results yield $C_{si} = 5.535$ and $Q_{ox} = 1.844$ in these cases. The percentage error in C_{si} and Q_{ox} is plotted in Figure 4. From these results it can be seen that the induced interfacial flux does not have much effect when the diffusivity ratio is near unity but as this ratio decreases errors of up to 20% occur if the interfacial flux is not included.

8. EXPERIMENTAL RESULTS FOR BORON DIFFUSION AND OXIDE REDISTRIBUTION

In the remaining of this paper we present an application of the numerical simulation described previously, in recent experimental study of boron diffusivity and segregation in intrinsic silicon (low dopped) and Si/SiO_2 interfaces respectively.

Boron is the dominant p-type dopant in silicon technology. However, even though it has been the subject of intensive study over the last two decades there still exists ambiguity in the above two essential parameters. A contributing factor to this ambiguity results from the inadequacy of simulation tools used to extract experimental results.

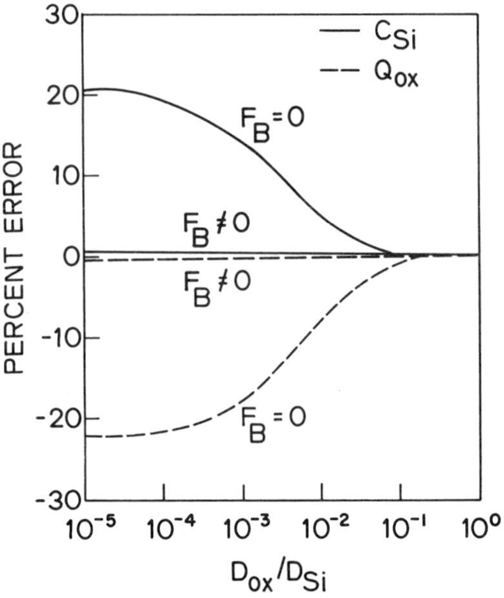

Figure 4. The percentage error in C_{si} and Q_{ox} plotted against the diffusivity ratio D_{ox}/D_{si}. The calculations are made for distributions found numerically with and without the interfacial flux, F.

It is presently established that boron diffuses at least partly by means of charged defect interactions [11]. For this reason the diffusivity of boron is affected by substitutional impurities in silicon, present at sufficient levels to cause the crystal to become extrinsic at the fabrication temperatures. The results of this effect are to a large degree responsible for the abundance of conflicting values of boron diffusion coefficients.

It has also been established that the diffusivity of boron in silicon is enhanced by the growth of silicon dioxide at the surface [12]. The degree of enhancement is greater in <100> crystal orientation than in <111>. Again a number of conflicting diffusion coefficients have been reported owing to the coexistence in experiments of extrinsic effects, oxidation effects and redistribution of boron at the moving silicon-silicon oxide boundary[13].

The redistribution of boron during oxidation is, to a large degree, controlled by the segregation (or partition) process at the moving silicon-oxide interface. This physical process when in thermodynamic equilibrium, gives rise to the so-called equilibrium segregation coefficient, m_{eq}, defined as the ratio of equilibrium impurity concentrations across the silicon-oxide interface ($m_{eq} \triangleq [C_{si}/C_{ox}]_{eq}$). Theoretical calculations only bracket the value of m_{eq} within several orders of magnitude. It is possible that during oxidation the segregation process may not be in equilibrium [14]. It thus follows that the actual impurity ratio, denoted here by m, might not be either equal to m_{eq} or even constant throughout an oxidation process step. A first order kinetic model can be easily incorporated to our algorithm. The basic assumption is that equilibrium of the segregation across the silicon-oxide interface is achieved in a flux limited manner expressed by

$$F_s = h_s(C_{ox} - C_{si}/m_{eq}) \tag{10}$$

where F_s is the non-equilibrium flux and h_s a kinetic limiting factor (velocity). Due to the motion of the interface during oxidation there also exists an induced flux, F_b across the interface given by

$$F_b = -V_{ox}\ (C_{ox} - \alpha C_{si}) \tag{11}$$

Thus, two competing fluxes exist at the silicon-oxide interface during oxidation. Thermodynamic equilibrium of the segregation process is approached only when $h_s >> V_{ox}$, a condition that might not be fulfilled for all the duration of an oxidation process since V_{ox} is inversely proportional to the square root of time. In terms of the algorithmic constraints discussed in section 4, the flux term F_s must always be considered irrespective of cell boundary crossing by the interface while F_b must be treated similarly to F_b in the above section.

The initial distribution of boron for all experiments was established by means of an ion implantation and annealing process that insures the intrinsic silicon requirements of the experiments.

For extracting the diffusivities and segregation coefficients we have used a new process simulator [6] that incorporates the previously described numerical technique, together with an automatic optimizer program that zeroes-in to the required parameters that yield the measured data. For the determination of diffusivities in inert atmosphere the only measured data were the initial and final sheet resistance for each experiment. Recently reported hole mobility data were used for the calculation of sheet resistances resulting from the simulation.* The reliability of this technique and the integrity of the mobility data is demonstrated by the excellent fit of the results by a simple activation energy and the agreement with other reliable data (Figure 5). Subsequent to the oxidation experiments both sheet resistances and juncton depths have been measured. The two process parameters thus determined by the simulation were the diffusivity and the segregation coefficient. In general, diffusivity enhancement has been obtained, more so for <100> than for <111>. These results are shown in Figure 6. For both orientations the enhancement increases as the temperature is decreased. The results cannot be represented by a single activation energy.

The segregation coefficients determined here under the assumption of thermal equilibrium (large h_s), are generally higher than those measured by Colby and Katz [14]. Also in all cases the calculated segregation coefficient is higher in <100> than in <111> silicon, but the difference is not as large as that reported by the above workers. In the present round of experiments there are insufficient data for the determination of h_s and m_{eq}. However, the consistent difference between directly

---

* Consult National Bureau of Standards

measured and calculated (equilibrium) m does point out to possible existence of non-equilibrium. Assuming the Colby and Katz data to represent near equilibrium conditions, values of h_s of the order of 10^{-3} μm/sec should be used to fit our experimental data.

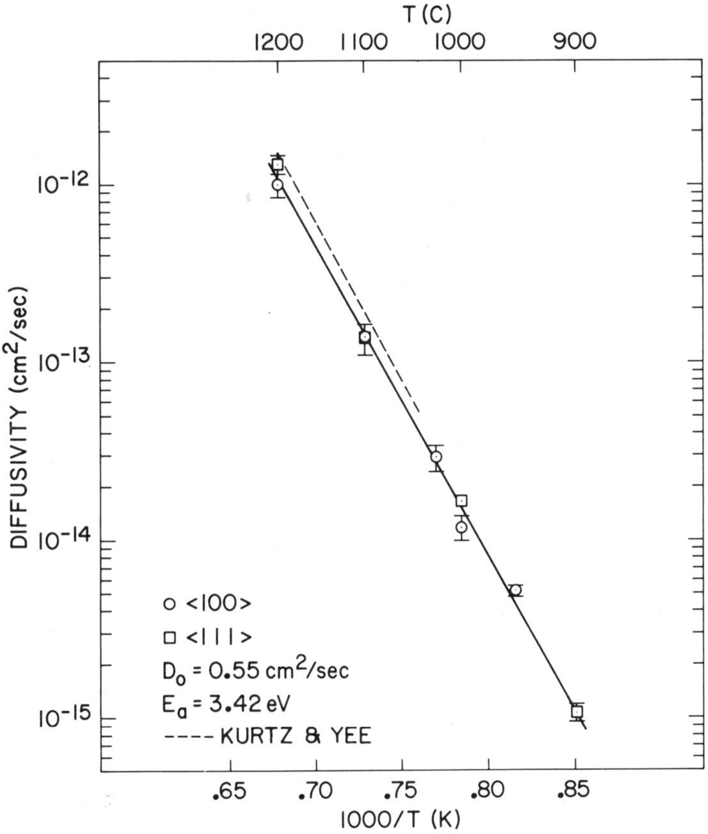

Figure 5. Boron diffusivity in intrinsic silicon in non-oxidizing ambient.

9. CONCLUSION

The oxidation of silicon and the resulting distribution of impurity dopants is the backbone of the silicon technology. Accurate numerical solutions to the moving boundary problem that arises requires that induced interfacial fluxes be explicitly

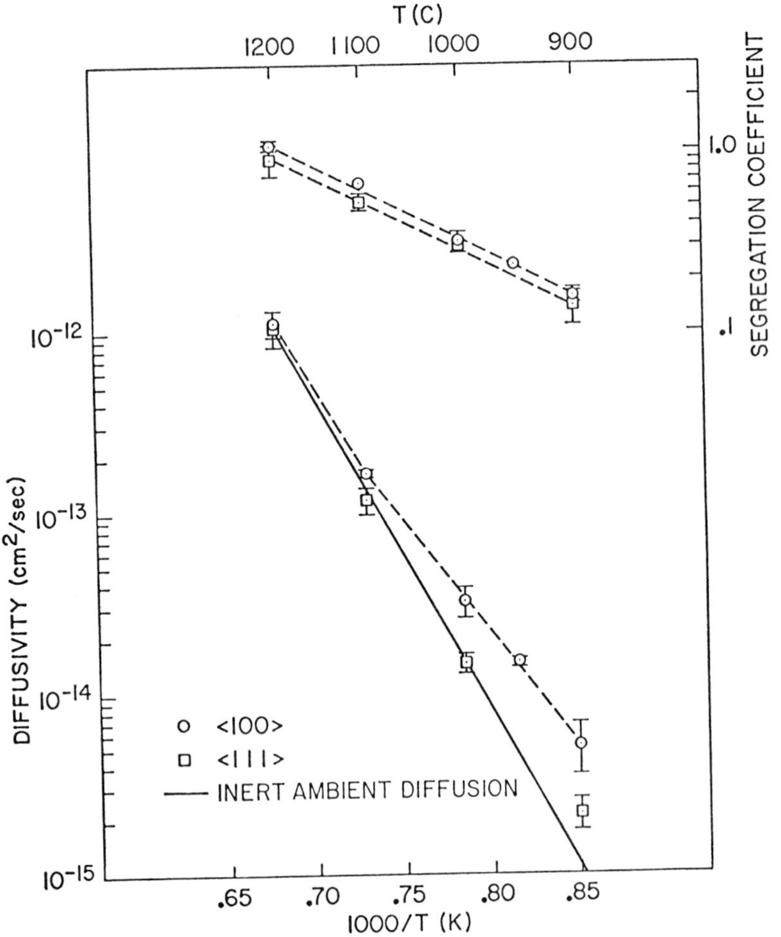

Figure 6. Boron diffusivity in silicon and segregation coefficient in an oxidizing ambient.

included when cell boundaries are crossed by the moving silicon-silicon oxide interface.

This term may not be included at every time step and its inclusion depends on the exact spatial and temporal discretizations chosen. However, if this term is neglected, significant errors in impurity distributions may occur.

The method presented here allows precise calculations of

distribution profiles for wide ranges of diffusivity ratio and segregation coefficient which is necessary if conditions found in actual silicon device fabrication are to be accurately modeled during oxidation.

One application of the numerical scheme presented here is the study of boron diffusion in low doped silicon. Both the diffusivity and the segregation coefficient have been experimentally determined. The results for diffusivity show crystal orientation dependent enhancement. A similar dependence is obtained for segregation coefficients. A simple model that may account for non equilibrium segregation has been discussed.

REFERENCES

1. A. Grove, O. Leistiko, Jr., and C. T. Sah, "Redistribution of Acceptance and Donor Impurities during Thermal Oxidation in Silicon", J. Appl. Phys., Vol. 35, p. 2695, 1964.

2. C. P. Wu, E. C. Douglas, and C. W. Mueller, "Redistribution of Ion-Implanted Impurities in Silicon During Diffusion in Oxidizing Ambients".

3. M. Av-Ron, M. Shtzkes, P. J. Burkhardt, and C. Cadoff "Distribution of Dopant in SiO_2-Si", J. Appl. Phys., Vo.. 47, No. 7, July 1976.

4. J. L. Prince and F. W. Schwettmann, "Diffusion of Boron From Implanted Sources Under Oxidizing Conditions", J. Electrochem Soc., Vol. 121, p. 705, 1974.

5. H. Guckel and L. A. Hall, "A treatment of Impurity Diffusion in Oxidizing Ambients." Solid State Electronics, Vol. 18 pp 99-104, 1975.

H. Guckel and L. A. Hall, "An Approximate Model for Boron Drive Diffusions on Oxidizing Ambients", Solid State Electronics, Vol. 18, pp 875-879, 1975.

6. D. A. Antoniadis, S. H. Hansen, R. W. Dutton and A. G. Gonzalez, "SUPREM 1 -- A Program for IC Process Modeling and Simulation, Stanford Electronics Laboratory Technial Report, SEL 77-006, May 1977.

7. F. F. Morehead Extended Abstracts, Vol. 74-2, Electrochem Soc. p. 474 (New York Meeting), October 1974.

8. R. Kraft, "Finite Difference Techniques for Diffusion and Redistribution Problem with Segregation-type Boundary Conditions" Advances in Computer Methods for Partial Differential Equations, June, 1975.

9. M. Rodoni, O. Buneman, and R. W. Dutton. "Boron Redistribution After Oxidation", Extended Abstracts, Vol. 76-2, p. 850, (Las Vegas Meeting 10/76).

10 . Henrici, Discrete Variable Methods in Ordinary Differential Equations, John Wiley & Sons, Inc., New York-London-Sydney, 1962.

11. B. L. Crowder, J. F. Ziegler, F. F. Morehead and G. W. Cole. "The Application of Ion Implantation to the Study of Diffusion of Boron in Silicon." Article from "Ion Implantation in Semiconductors and Other Materials," pp. 267-274. Edited by Billey L. Crowder, Plenum Publishing Corp.

12. W. G. Allen. "Effect of Oxidation on Orientation Dependent Boron Diffusion in Silicon, Solid State Elec., Vol. 16, 1973, pp. 709-717.

13. W. G. Allen and C. Atkinson. "Comparison of Models for Redistribution of Dopants in Silicon During Thermal Oxidation, Solid State Elec., Vol. 16, 1973, pp. 1283-1287.

14, J. W. Colby and L. E. Katz. "Boron Segregation at Si-SiO_2 Interface as a Function of Temperature and Orientation," J. Elec. Soc., Vol. 123, No. 3, March 1976, pp. 409-412.

Robert W. Dutton , Dimitri A. Antoniadis
Integrated Circuits Laboratory
Stanford Electronics Laboratories
Stanford University
Stanford, CA 94305

This research was sponsored under ARO contract DAAG-29-77-C-006
The thesis contribution of Madeline Rodoni and Adalberto Gonzalez are the core of materials presented here.

MOVING BOUNDARY PROBLEMS

LOW-SPEED DIPFORMING

Michael F. Malone and Gabriel Horvay

Dipforming is a continuous cladding process used to deposit a uniform layer of molten material on a cold bar which plays the role of a heat sink. For technological reasons the input bar enters the bath of molten material from below. The controlling step in the process is one of heat transfer. An energy balance alone determines the position of the interface, as well as the temperature variation in the bar.

The non-linear partial differential equation describing the process can be approximated with a system of two non-linear ordinary differential equations [1]. The equations are coupled, but first order when axial conduction is ignored.

This work considers the second order problem, including axial conduction. Understanding of this aspect requires consideration of the three-region problem, which examines behavior in the molten bath and in the sections to its "left" (i.e., below the bath), and to its "right" (i.e., above the bath). Boundary conditions for the problem are not obvious, but are similar to conditions used in the analysis of some chemical reactor problems [3].

The Peclet number (the speed parameter) for dipforming is typically large, p=O(10). Since ε, the squared reciprocal of the Peclet number appears as a parameter premultiplying the highest order derivative, a solution is sought using the method of matched asymptotic expansions.

The expected boundary layer behavior near the bath entrance is observed. However, near the exit, special care must be taken in constructing a uniformly valid solution. The solution obtained qualitatively explains discrepancies between predictions from [1] and experiments for copper-clad copper. Physical explanation of the behavior at the exit is given, and directions for additional work are also suggested.

1. INTRODUCTION

Dipforming refers to a process whereby a cold input bar of material I is drawn through a crucible containing a bath of molten material M. The entrance radius of the rod (or half-width of the slab, for convenience also to be referred to as radius) X_e is first enlarged as a layer of M freezes out onto I. The bar will

ISBN 0-12-757350-X

emerge with increased radius X_E when the speed of traverse is high and the bath not too deep. Lower speeds or deeper baths result in a radius decrease or meltback. Complete meltback, or melt-out, occurs for low speeds of the bar or tall crucibles. In practice, the traverse proceeds upward, to avoid bowing due to gravity (that would occur in a horizontal assembly), and to avoid difficulties that would be created by a submerged exit port. Some of the technological difficulties created at the submerged entrance port (the need to maintain a proper seal, and the need to avoid seizing at the entrance) are discussed in [1]. Fig. 1 illustrates the qualitative behavior of a bar in the dipforming process. Axial conduction is expected to produce a situation similar to that of Fig. 2. (In these figures the entrance radius is denoted by X_o. Figures 1 and 2 are reproduced from [1] by permission of ASME.) We expect axial conduction to increase pickup [$(X_E-X_e)/X_e$ for slab geometry, and $(X_E-X_e)^2/X_e{}^2$ for rod geometry] and melt-out time.

Commercial processes often clad materials which have good electrical conductivity (e.g. copper) onto input bars with good mechanical strength (e.g. steel). But copper onto copper dipforming, in producing electrical cable of suitable diameter, is probably the most important commercial operation. (See e.g. [2].) We examine here the case where the input bar I and the melt M are of identical materials, and for mathematical simplicity we also choose a slab geometry instead of the rod geometry. We shall also use an approximation which retains only the average temperature and the radius as dependent variables. The rod geometry, with calculation of temperature profiles across the bar is more important technologically, but introduces complications which obscure the essential features of the present analysis.

In Section 2 the problem is described and the governing equations are introduced. Boundary conditions on temperature are discussed and compared with the classic problem of axial dispersion in a chemical reactor [3]. The solution ignoring axial conduction and motivation for refinement appear in Section 3. In

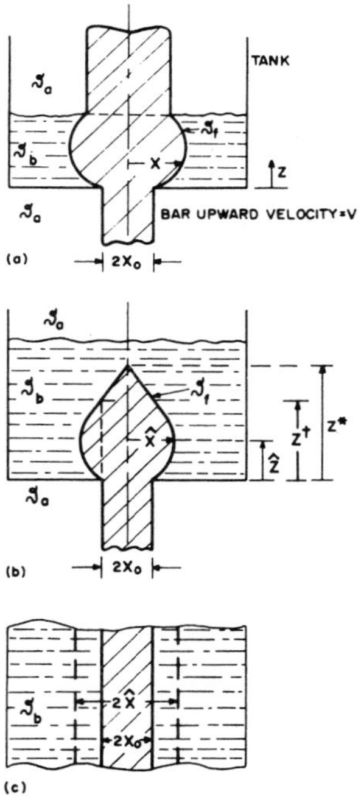

Figure 1. Illustration of a cold metal bar which enters, at speed V, a tank containing a bath of molten metal. (a) The bath is shallow; (b) the bath is deep; (c) the slab velocity is infinite (the case of sudden immersion).

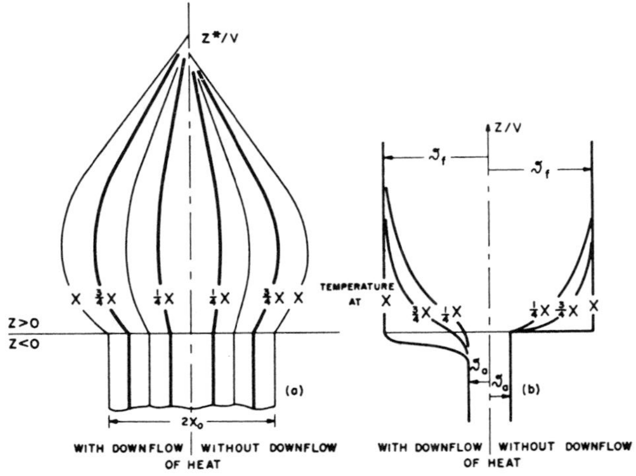

Figure 2. The case of slow slab motion (left half of diagrams) is contrasted with fast slab motion (right half of diagrams) for the two-channel model. Diagram (a) illustrates the growth, then contraction of the slab radius X. Diagram (b) gives a schematic sketch of the temperature variation at locations X, 3X/4, X/4. When the slab moves slowly, heat seeps out into the $Z < 0$ region, and the melt-out time is increased.

Section 4 a solution of the 3-region problem is obtained, and its behavior near the entrance and near the exit is discussed. Finally, in Section 5 a physical explanation of the behavior at the exit is given, and directions for further work are suggested, along with discussion of a representative example.

2. PROBLEM STATEMENT

Reference [1] established the governing partial differential equation for the bath section, and constructed an approximation of the partial differential equation consisting of two non-linear ordinary differential equations. A summary of the equations for the middle section M and appropriate equations in the regions to the left (L) and to the right (R) of the bath are presented below. Fig. 1 sketches the system we shall consider.

Two nodal lines are considered, one at the interface, and another at the center of the bar half-width (one-channel model). We retain a continuous representation in the axial direction. Convective transfer occurs at the interface, with constant heat transfer coefficient h in the bath and h_L and h_R to the left and to the right of M. Radial conduction occurs from the interface node (which is at the fusion temperature of the melt) to the internal node. The surface node gains heat from the bath superheat, from the release of latent heat of fusion, and due to mass transfer from bath to bar. A more detailed discussion of the physics appears in [1].

For both the left ($Z<0$) and the right ($Z>Z_E$) regions the following dimensionless equation holds for the average temperature U:

$$\varepsilon\ddot{U}-\dot{U}-2U = -2u_a \tag{1}$$

For simplicity we have assumed infinite values for h_L, h_R, so that the surface of the bar is isothermal, at the ambient temperature below and above the bath. (1) is obtained from (4b) of [1] by deleting terms due to phase change, and replacing the surface temperature by the ambient value.

In the bath section, energy balances at the surface node and the internal node result in the following expressions

surface node ϑ_f: $(1+U)\dot{R}-2U/R+u_B = 0$ (2a,b)

internal node ϑ: $R\dot{U}+2U/R = \varepsilon(R\dot{U})^{\cdot}$

We have used the dimensionless notation

$$R = X/X_e,\ z = Z/X_e,\ \tau = t/(X_e^{\,2}/\kappa),\ p = X_eV/\kappa,\ \mathcal{B} = hX_e/k \quad (3a)$$

$$u_a = c(\vartheta_f-\vartheta_a)/\lambda,\ U = c(\vartheta_f-\vartheta)/\lambda,\ u_B = \mathcal{B}c(\vartheta_b-\vartheta_f)/\lambda \quad (3b)$$

for slab radius, axial distance, time (Fourier number), speed of travel of the slab (Peclet number), heat transfer coefficient from melt to slab (Biot number), ambient temperature, temperature at the internal node (both positive when ϑ_a, ϑ are below the fusion temperature ϑ_f), and effective superheat temperature (positive when the bath temperature ϑ_b is higher than ϑ_f).

We really solve a <u>steady state problem</u>, inasmuch as a coordinate system fixed in the laboratory is being considered. The notation τ refers to time from entry into the bath. For the assumed steady velocity V of travel we have

$$\tau = t/(X_e^{\,2}/\kappa) = z/p \quad (3c)$$

In the above a dot indicates τ derivative, X is the radius, Z is axial distance from the entrance, k is thermal conductivity, κ thermal diffusivity, c specific heat, λ heat of fusion. ε, the small perturbation parameter arising in the axial conduction term, is the reciprocal of the squared Peclet number:

$$\varepsilon = 1/p^2 \quad (3d)$$

At the time of the publication of [1] Peclet numbers of the order of 5 were prevalent, nowadays $p \sim 20$ ($V \sim 1$ meter/sec) are more representative values. Nevertheless, the $p = \infty$ solutions lead to results which reveal a departure from experimental observations. Moreover, relatively low-speed operations are, for cases where thermal shock must be avoided, not without interest per se.

With this motivation we propose to obtain a perturbation

solution of (1) and (2).

We note that equations slightly different from (2) also constitute a valid approximation for the problem (interpreting material pickup at the interface in a slightly different manner), and are also presented in [1]. The system (2) is somewhat more convenient and will be used here. For a discussion of the physical differences between the two approximations see [1] and [4].

Boundary conditions for the radius are obvious. From the definition in (3a)

$$\tau < 0: \quad R=1 \tag{4}$$

i.e., the width is constant below the entrance, and at the entrance we have

$$\tau = 0: \quad R(0) \equiv R_e = 1 . \tag{5}$$

At the bath exit on the right, specified by τ_E, the radius again attains a constant value

$$\tau = \tau_E: \quad R(\tau_E) = R_E \tag{6}$$

A suitably large R_E is the objective of commercial practice. However, if the bath is deep enough, or the speed slow enough, melt-out will occur. We define the smallest such time as the melt-out time

$$\tau_E = \tau^*: \quad R^* = R(\tau^*) = 0 \tag{7}$$

Boundary conditions for temperature are analogous to the conditions encountered for the dependent variable in the analysis of chemical reactors with axial disperson. For the chemical reactor reactants (mass) rather than heat are transported axially, and concentration, rather than temperature, is the dependent variable.

Recall that (1) holds in the regions L and R (we label the variables R_L, ... , U_R while in the middle section we write R,U without subscripts). We begin with the intuitive notion that the temperatures must be continuous in passing between regions, and obtain

$$U_L(0) = U(0) \equiv U_e \tag{8}$$

$$U_R(\tau_E) = U(\tau_E) \equiv U_E \tag{9}$$

Energy is transported axially both by convection and by conduction. (For the chemical reactor, mass is transported both by convection, and a mechanism analogous to conduction of heat which may include both molecular diffusion and hydrodynamic effects.) Energy balances at the entrance (heat convected in, minus heat conducted out) and exit result in the relations

$$U_L(0) - \varepsilon_L \dot{U}_L(0) = U(0) - \varepsilon \dot{U}(0) \tag{10}$$

$$U(\tau_E) - \varepsilon \dot{U}(\tau_E) = U_R(\tau_E) - \varepsilon_R \dot{U}_R(\tau_E) \tag{11}$$

The Peclet number for our case (I = M) will be the same for all three sections, and subscripts on ε in (10) and (11) may be dropped. For $I \neq M$ an effective Peclet number for each section may be different, and may lead to a discontinuous slope in U at the junction. A discussion of this point for the chemical reactor problem appears in [3].

With ε a constant for the entire region $-\infty < \tau < \infty$ we obtain, using (8) with (10) and (9) with (11)

$$\dot{U}_L(0) = \dot{U}(0) \equiv \dot{U}_e \tag{12}$$

$$\dot{U}_R(\tau_E) = \dot{U}(\tau_E) \equiv \dot{U}_E \tag{13}$$

Isothermal conditions at the surface in the L and R regions lead to the final two conditions required for U:

$$U_L(-\infty) - \varepsilon \dot{U}_L(-\infty) = U_L(-\infty) = u_a \tag{14}$$

$$U_R(\infty) + \varepsilon \dot{U}_R(\infty) = U_R(\infty) = u_a \tag{15}$$

(1) applied to U_L and U_R with (2) subject to (5), (8), (9), (12) - (15) define the problem at hand.

3. SOLUTION FOR THE MIDDLE REGION WHEN MELT-OUT IS REACHED IN THE BATH. A NOT SO SINGULAR PERTURBATION PROBLEM.

In [1] only the solution of (2) subject to

$$\varepsilon = 0: \quad U(0) = U_e, \quad U_e = u_a \tag{16a,b}$$

and (5) was considered. In fact, for $\varepsilon = 0$ the complementary condition

$$\dot{U}(0) = \dot{U}_e \tag{16c}$$

is dropped since the problem becomes first order. Actually, as will be seen, $\dot{U}_e$ is determined by (12).

Before obtaining the $\varepsilon = 0$ solution we note that on adding (2a) and (2b) we obtain

$$[R(1+U-\varepsilon\dot{U})]^{\cdot} + u_B = 0 \tag{17}$$

which, in the light of (5), integrates into

$$R(1+U-\varepsilon\dot{U}) = u_B(\tau^*-\tau) \tag{18}$$

($u_B\tau^*$ is the integration constant) and in accordance with the definition

$$R(\tau^*) = 0 \tag{19a}$$

specifies the melt-out time as

$$\tau^* = [1+U_e-\varepsilon\dot{U}_e]/u_B \tag{19b}$$

Writing the perturbation expansions

$$R = r + \varepsilon s, \quad U = u + \varepsilon v \tag{20}$$

the equations (2a,b), (18) become the zeroth order set

$$\varepsilon^\circ: \quad \begin{aligned} &(1+u)\dot{r} - 2u/r + u_B = 0 \\ &r\dot{u} + 2u/r = 0 \\ &r(1+u) = u_B(\tilde{\tau}-\tau) \end{aligned} \tag{21a,b,c}$$

and the first order set

$$\varepsilon^1: \quad \begin{aligned} &(1+u)\dot{s}-2v/r+u_B s/r+v\dot{r}+(1+u)s\dot{r}/r = 0 \\ &r\dot{v}+2v/r+2s\dot{u} = \ddot{u}+\dot{u}\dot{r}/r \\ &(1+u)s+vr-r\dot{u} = (1+u_e)s_e+v_e-\dot{u}_e \end{aligned} \qquad (22a,b,c)$$

where the zeroth order approximation to the melt-out time

$$\tilde{\tau} = (1+u_e)/u_B \qquad (23)$$

is smaller than the rigorous value (19b) because of the absence of the conduction loss term, $-\varepsilon\dot{u}_e$.

The equations (21c) and (22c) are redundant; any two of the three equations in each set may be chosen as governing.

The boundary conditions for the zero order set are the entrance conditions

$$r(0) = 1, \; u(0) = u_e, \; -\varepsilon\dot{u}(0) = -\varepsilon\dot{u}_e = u_a-u_e \qquad (24a,b,c)$$

and the melt-out conditions

$$r(\tilde{\tau}) = 0, \; u(\tilde{\tau}) = 0, \; \dot{u}(\tilde{\tau}) = 0 \qquad (24d,e,f)$$

The right side of (24c) follows from (10) in the light of the assumption that U_L = const = u_a (one-region problem). Note that $-\varepsilon u_e$ is positive (heat is conducted out), and is really not a completely negligible quantity when ε is not precisely zero.

By (21b)

$$r^2 = 2u/(-\dot{u}) \qquad (25)$$

In conjunction with (24) this leads to

$$-\dot{u}_e = 2u_e, \quad u_e = u_a/(1+2\varepsilon) \qquad (26a,b)$$

Combining (25) and (21c) we obtain the implicit expression for u:

$$\frac{u}{1+u}\exp\left(\frac{1}{1+u}\right) = C\exp[-2/u_B{}^2(\tilde{\tau}-\tau)] \qquad (27a)$$

where, by (16)

$$C = \frac{u_e}{1+u_e} \exp\left[\frac{2}{1+u_e} + \frac{2}{u_B{}^2\tilde{\tau}}\right] \tag{27b}$$

Fig. 3 plots r and u vs. τ for the parameter values

$$u_B = .4, \quad u_e = 5/3 \tag{28a}$$

Similar curves, obtained by an analog computer, are shown in [1] Fig. 4b, for $u_B = .4$, $u_e \equiv u_a = 2.0$.

Since the zeroth order approximations, r, u, fortuitously satisfy all boundary conditions (24a,b,c,d,e,f), there is no singular perturbation problem on hand; the melt-out condition (24d) eliminated a possibly singular behavior at the right end. However, one may seek improvement over the r,u solution, by determining the first order <u>ordinary perturbation</u> correction terms s,v from (22) for the boundary conditions

$$s(0) = s_e = 0, \quad v(0) = v_e = 0, \quad v(0) = v_e = 0 \tag{29a,b,c}$$

$$s(\tau^*) = 0, \; v(\tau^*) = 0, \; \dot{v}(\tau^*) = 0 \tag{29d,e,f}$$

From (22c) place

$$s = [r(\dot{u}-v)-\dot{u}_e]/(1+u) \tag{30}$$

into (22b). There results

$$\dot{v}+f(\tau)v = g(\tau) \tag{31a}$$

where

$$f(\tau) = \frac{2}{r^2} - \frac{2\dot{u}}{(1+u)r}, \quad g(\tau) = \frac{\dot{r}\dot{u}+\ddot{u}}{r^2} - \frac{2\dot{u}}{1+u} \tag{31b,c}$$

As well-known, the solution is

$$v = e^{-h(\tau)}\left[C + \int_0^\tau e^{h(\xi)} g(\xi)\,d\xi\right] \tag{32a}$$

where

$$C = 0 \tag{32b}$$

because of (29b). In the foregoing

$$h(\tau) = \int_0^\tau f(\xi)d\xi = \ell n\left[\left(\frac{1+u_e}{1+u}\right)^2 \left(\frac{u_e}{u}\right)\right] \tag{32c}$$

Hence

$$v(\tau) = \left[\frac{1+u(\tau)}{1+u_e}\right]^2 \left[\frac{u(\tau)}{u_e}\right] \int_0^\tau \left[\frac{1+u_e}{1+u(\xi)}\right]^2 \left[\frac{u_e}{u(\xi)}\right] g(\xi)d\xi \tag{33}$$

Numerical integration leads to the curves $R \simeq r+\varepsilon s$, $U \simeq u+\varepsilon v$ also plotted in Fig. 3 for

$$\varepsilon = .1, \quad u_a = (1+.2)u_e = 2.0 \tag{28b}$$

Note that the r, u, v curves are defined as zero beyond $\tilde{\tau}$, whereas the $r+\varepsilon s$ curve terminates at τ^*, passing, by (30), through the point $-\varepsilon\dot{u}_e$ at $\tilde{\tau}$:

$$\tilde{\tau}<\tau<\tau^*: \quad R = r+\varepsilon s = -u_B(\tau-\tilde{\tau})-\varepsilon\dot{u}_e = u_B(\tau^*-\tau) \tag{34}$$

(linear variation). Note the shift to the right of the R curve compared to the zeroth order curve r.

When the full 3-region problem is considered, we see that, in the absence of melt-out, the $\varepsilon = 0$ solution cannot satisfy all of the boundary conditions imposed on U in (8), (9), (12), (13); this will be shown quantitatively in Section 4. Comparison of solutions similar to (2) for the rod case with experiments (see [1]) agree well only when the predicted values for R are shifted to the right by a small amount and are raised somewhat (approximately 10% when $\varepsilon \sim .04$). Alternately, an increase of about 10% in u_a also brings the theory and experiment into better agreement. Thus, we have both analytical and experimental motivation to examine the effect of finite, though small, values of ε.

4. SOLUTION OF THE 3-REGION PROBLEM FOR FINITE ε

Exact solution of (1) for both L and R regions is possible and may be written as

$$U_k = u_a+K_{1,k}e^{\lambda_1\tau}+K_{2,k}e^{\lambda_2\tau} \qquad k = L,R \tag{35}$$

with

$$\lambda_1 = [1-\sqrt{1+8\varepsilon}]/2\varepsilon, \quad \lambda_2 = [1+\sqrt{1+8\varepsilon}]/2\varepsilon \tag{36}$$

independently of the coupling to the M solution. Note that $\lambda_1<0$, $\lambda_2>0$. In accordance with (14), (15) $K_{1,L} = K_{2,R} = 0$, and therefore

$$U_L = u_a+K_L e^{\lambda_2\tau}$$
$$U_R = u_a+K_R e^{\lambda_1\tau} \tag{37a,b}$$

Determination of the middle solution M must be carried out with approximate methods since exact closed form solution of (2) for finite ε is impossible. As in Section 3 we proceed to find the outer solution ignoring ε terms and obtain nearly identical results to the previous section. Since the coupled problem is now considered, (16b) no longer holds.

Application of (8) and (12) leads to the expressions for u_e and K_L:

$$u_e = \frac{\lambda_2}{2+\lambda_2} u_a, \quad K_L = -\frac{2}{2+\lambda_2} u_a \tag{38a,b}$$

Furthermore by (25)

$$\dot{u}_e = -2u_e \tag{38c}$$

holds. (Note that for $\varepsilon<<1$, (38a) reduces to (26b).)

For $0<\varepsilon<<1$ a boundary layer is seen to exist near $\tau = 0$ where the L solution undergoes a rapid decrease due to axial conduction. (A rigorous solution of this problem, in the absence of freezing in M, for arbitrary heat transfer coefficients in L and M, was given, via the Wiener-Hopf technique, in [5,6].) For decreasing ε the transition becomes very sharp, but all boundary conditions may still be satisfied, contrary to the solution obtained in Section 3. Use of (23) underestimates τ^*. For finite ε, τ^* is given by (19b), with u_e as in (38a) and $\dot{u}_e$

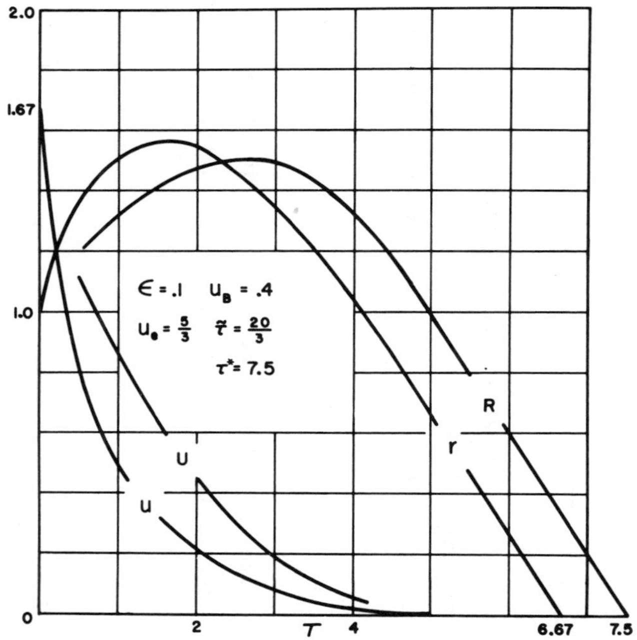

Figure 3. Middle region inner solution in zereth (r,u) and first order (R,U) approximations

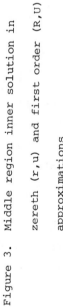

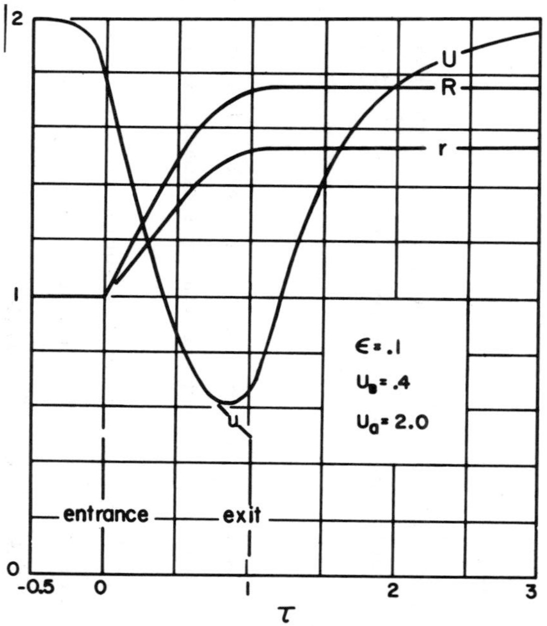

Figure 4. Uniformly valid zereth approximation (R,U) to the solution of the 3-region problem. The zereth order inner solution r,u is also shown

as in (38c). Correspondingly,

$$\tau^* = \underset{\sim}{\tau}^* \equiv [1 + \frac{\lambda_2}{2+\lambda_2} (1+2\varepsilon)u_a]/u_B \tag{39}$$

Note that the solution in region L could have been obtained approximately by singular perturbation technique, in the form

$$U_L = u_a + K_L e^{+\tau/\varepsilon} \tag{40}$$

valid for small ε, since $\lambda_2 \to 1/\varepsilon$ as $\varepsilon \to 0$. For the R solution singular perturbation techniques are not suitable inasmuch as $\lambda_1 \to -2$ as $\varepsilon \to 0$.

At the bath exit, τ_E, the outer M solution terminates at a value u_E given implicitly by (27) with $\tilde{\tau}$ replaced by $\underset{\sim}{\tau}^*$ and C of (27b) redefined with $\underset{\sim}{\tau}^*$ as $\underset{\sim}{C}$, eqs (27a,b). K_R in (37b) can be determined from (9) if the outer solution is used, as

$$K_R = (u_E - u_a)e^{-\lambda_1 \tau_E} \tag{41}$$

Hence

$$U_R = u_a + (u_E - u_a)e^{\lambda_1(\tau - \tau_E)} \tag{42}$$

But (13) cannot in general be satisfied for finite ε, and we are led to seek an improvement in u.

We seek an asymptotic expansion in the boundary layer variable

$$\theta = (\tau_E - \tau)/\varepsilon \tag{43}$$

In terms of θ, (2) gives rise to the zeroth order "inner" equations (prime denotes θ derivative; we shall use script letter $\mathcal{R},\mathcal{U}$ for inner solution)

$$\begin{aligned} &(1+\mathcal{U})\mathcal{R}' = 0 \\ &\mathcal{R}\mathcal{U}' + (\mathcal{R}\mathcal{U}')' = 0 \\ &[\mathcal{R}(1+\mathcal{U}+\mathcal{U}')]' = 0 \end{aligned} \tag{44a,b,c}$$

and the solution is

$$\mathcal{R}' = 0, \quad \mathcal{U} = C_1 + C_2 e^{-\theta} \tag{45}$$

Note that (44a) merely states that $\mathcal{R}' = 0$ to the extent that ε terms are neglected. A more suitable estimate for $\mathcal{R}$ is obtained using (18). The "inner" solution given by (45) is valid inside the boundary layer, and is used with (9) and (13) to determine C_2 and K_R in terms of C_1:

$$C_2 = \frac{\varepsilon\lambda_1}{\varepsilon\lambda_1 - 1}(u_a - C_1), \qquad K_R = \frac{e^{-\lambda_1 \tau_E}}{\varepsilon\lambda_1 - 1}(u_a - C_1) \tag{46a,b}$$

The conventional matching procedure (see e.g. [7], [8]) determines C_1. This matching procedure requires that the outer limit of the inner solution equal the inner limit of the outer solution, or more formally, following [8], p. 297: for fixed Θ,

$$\lim_{\varepsilon \downarrow 0} [u(\tau)]_{\tau = \tau_E - \Theta\delta} = \lim_{\varepsilon \downarrow 0} [\mathcal{U}(\theta)]_{\theta = \Theta\delta/\varepsilon} \tag{47}$$

Here $\delta(\varepsilon)$ is some suitable function of ε which satisfies

$$\lim_{\varepsilon \downarrow 0} \delta = 0, \qquad \lim_{\varepsilon \downarrow 0} \delta/\varepsilon = \infty \tag{48a,b}$$

So (47) gives

$$C_1 = u(\tau_E) \tag{49}$$

To obtain a uniform solution in M, we add the inner and outer solutions and subtract the common part:

$$U \simeq u + C_2 e^{-(\tau_E - \tau)/\varepsilon} \tag{50}$$

and our approximate solution is now determined.

5. CONCLUSIONS AND DISCUSSION

Fig. 4 plots R and U as functions of τ. We have chosen the representative values

$$\varepsilon = .10,\ u_B = .4,\ u_a = 2.0,\ \tau_E = 1.025 \text{ (corresponding to } u_E = .5) \tag{51}$$

The plots clearly show the boundary layers at the entrance $\tau = 0$ and at the exit τ_E. The outer M solutions, r,u are also shown. Note the increase in R contrasted with r, in qualitative agreement with the experiments cited in [1]. (Recall that the experiments in [1] pertain to rod geometry.)

Boundary layer behavior in the U_R solution does not appear. Physically, the reason is clear from an examination of Fig. 4. Remember that U is defined as positive, and rising temperature ϑ corresponds to decreasing U. At the entrance heat is convected in the positive τ direction and conducted in the opposite direction. These opposing modes of transport create the boundary layer near $\tau = 0$, since for large speeds (small ε) convection dominates. Near τ_E, however, both convection and conduction occur in the positive τ direction, and though convection may dominate, the modes of transport do not offset, but complement each other.

Generalization of this work to include several additional features is suggested. (A) Extension to rod geometry and determination of radial temperature profiles using a multichannel model (to permit assessment of thermal stresses) is of considerable manufacturing importance, and is of first priority. (B) The assumption of having the same ambient temperature in L and R may not be realistic, nor having isothermal conditions in L and R. Removal of these limitations should be the subsequent task. (C) The problem for I different from M can also be handled using singular perturbation techniques, and would constitute a further generalization of the present analysis.

REFERENCES

1. G. Horvay, "The Dipforming Process," J. Heat Transfer, vol. 87, 1965, p. 1.

2. Dip-Forming Process, 1977 brochure of the General Electric Co., Wire and Cable Division, Advanced Technology Operation, Bridgeport, CT 06602.

3. J. F. Wehner and P. H. Wilhelm, Chem. Eng. Sci., vol. 6, 1956, p. 89.

4. G. Horvay and I. Giaever, "Cladding by Dipforming," Proc. 1967 Symp. of Jap. Soc. Mech. Eng., Tokyo.

5. G. Horvay, "Temperature Distribution in a Slab Moving from a Chamber at One Temperature to a Chamber at Another Temperature," J. Heat Transfer, Vol. 83, 1961, p. 391.

6. G. Horvay, "The Effect of Discontinuous Biot Number on the Temperature Distribution in Moving Slabs," Proc. Internatl. H.T. Conf. at Boulder, ASME 1961, p. 7.

7. J. Cole, Perturbation Methods in Applied Mathematics, Blaisdell, 1968.

8. C. C. Lin and L. A. Segel, Mathematics Applied to Deterministic Problems in the Applied Sciences, MacMillan, 1974.

Michael F. Malone, Department of Chemical Engineering
Gabriel Horvay, Department of Civil Engineering
University of Massachusetts, Amherst

ACKNOWLEDGMENT

Partial support for this work with a University of Massachusetts Faculty Research Grant is acknowledged with appreciation. G. H. also wishes to express his indebtedness to the General Electric Co. for more than 20 years of wonderful association, with personnel and personal relations and scientific stimulus always at their best.

PERMAFROST THERMAL DESIGN FOR THE TRANS-ALASKA PIPELINE

J. A. Wheeler

Design of the Trans-Alaska pipeline required extensive analysis of the thermal regime in permafrost. A 2-D simulator based on variational methods was developed to perform these analyses. About 330 miles of the pipeline is supported above ground by thermally active support members designed with the simulator. These support members utilize heat pipes to maintain or produce a frozen foundation. Mechanical refrigeration was also used to maintain a frozen foundation.

1. INTRODUCTION

Permafrost is any soil or rock which has existed at a temperature colder than 32°F for two or more years [9]. It occurs extensively in arctic and subarctic regions; 82% of Alaska, 50% of Canada, and 50% of the USSR are underlain by permafrost. Usually permafrost is bounded above by a one to ten foot thick active layer which thaws each summer and refreezes each winter. Its lower boundary is defined by a relatively stationary 32°F isotherm at depths ranging from 20 to 2300 feet. In northern Alaska the permafrost is continuous and has a mean temperature of about 15°F near the earth's surface; in central Alaska it is discontinuous and has a mean temperature of about 31°F.

A high ice content in most permafrost soils complicates the design of pipelines and other facilities. This ice can be uniformly distributed in the soil or it can occur in massive layers or wedges of nearly pure ice. So long as the ice remains frozen it provides a firm foundation but if it thaws then differential subsidence, shear failure, or thermal erosion can destroy the usefulness of a structure. Foundation failure due

ISBN 0-12-757350-X

to thawing permafrost is usually a slow process occurring over a period of years.

Prevention of permafrost thaw below an oil pipeline is not a simple task. Chilling Prudhoe Bay crude before transmission is impractical because of its high viscosity at low temperatures; the oil temperature at design capacity will be about 145°F. Some of the Alyeska line in northern Alaska is insulated and buried but this mode of construction is expensive due to the large volume of insulation required. In central Alaska where natural permafrost temperatures are closer to 32°F, the volume of insulation required is completely impractical. About 420 miles of the Trans-Alaska pipeline is elevated above ground. This effectively separates the hot oil from the permafrost but still does not entirely prevent thawing. In central Alaska, construction disturbances at the soil-atmosphere interface tend to initiate thawing and thus some additional measure was required to maintain frozen soil around structural members supporting the elevated line.

2. EQUATIONS DEFINING HEAT TRANSFER IN PERMAFROST

Conduction with a change of state, Stefan's problem, is the principal thermal process of interest in permafrost. The process is complicated by unusual thermal properties and complex boundary conditions. Only the basic equations necessary for a simulator will be developed here. A review of the physics involved can be found in a paper by Anderson and Morgenstern [1].

Heat conduction with a change of state in an open region R with boundary σ can be formulated mathematically as follows:

$$L(T) = \frac{\partial}{\partial t}(\alpha T + s) - \vec{\nabla} \cdot K\vec{\nabla}T = 0; \quad \vec{X} \in R \tag{2.1}$$

$$T = T_a(\vec{X}, t) \quad ; \quad \vec{X} \in \sigma_a \tag{2.2}$$

$$\vec{\eta} \cdot \vec{q} = q_b \ (\vec{X},t,T) \quad ; \quad \vec{X} \ \varepsilon \ \sigma_b \tag{2.3}$$

$$T(\vec{X},0) = T^0 \ (\vec{X}) \quad ; \quad \vec{X} \ \varepsilon \ R \cup \sigma \tag{2.4}$$

Equation 2.1 represents an energy balance. Equations 2.2 and 2.3 represent boundary conditions and Equation 2.4 specifies the initial condition. The boundary σ is divided into two parts, σ_a and σ_b, according to the type of boundary condition imposed at any given time. Heat flux is defined by $\vec{q} = -K\vec{\nabla} T$ and $\vec{\eta}$ is a normal unit vector directed into the region R. Since the properties α, s, and K as well as the gradient of the temperature T are discontinuous at the melt front, Equation 2.1 can be more rigorously expressed as separate equations for each subregion in which continuity exists. This would also require a moving boundary condition at the melt front to couple the subregions. In the present context, Equation 2.1 provides a simpler notation and does not lead to errors.

The volumetric heat capacity α and the thermal conductivity K are defined by:

$$\alpha(\vec{X},T) = \begin{cases} \alpha_s \ (\vec{X}) \ ; \ T(\vec{X},t) < T_f \\ \alpha_L \ (\vec{X}) \ ; \ T(\vec{X},t) > T_f \end{cases} \tag{2.5}$$

and

$$K(\vec{X},T) = \begin{cases} K_s \ (\vec{X}) \ ; \ T(\vec{X},t) < T_f \\ K_L \ (\vec{X}) \ ; \ T(\vec{X},t) > T_f \end{cases} \tag{2.6}$$

where the nominal freezing point T_f is the highest temperature at which ice can exist in a given soil. Heat capacity and conductivity can be more generally represented as functions of temperature in frozen and thawed soils but this variation is usually small and almost invariably unknown for a given soil.

Pure water dispersed in a fine grained soil freezes over a finite temperature interval rather than at precisely 32°F. A fine soil such as a clay may have a specific surface area of up to 800 sq m/g. Interaction between water dispersed in a soil and the surface of the soil matrix alters the properties of the water. The net result is that a fraction of the dispersed water far from any soil surface freezes at 32°F and the remaining fraction freezes at some temperature less than 32°F. Typical curves relating unfrozen moisture content to temperature are shown in Figure 1. Care should be exercised in employing published data for unfrozen moisture because soils with the same classification may have very different properties and experimental determination of unfrozen water is difficult. In designing the Trans-Alaska pipeline we modeled unfrozen moisture with the expression

$$W_u(\vec{X},T) = \gamma(\vec{X}) \left[\frac{\alpha_u(\vec{X})}{\alpha_u(\vec{X}) + T_f - T(\vec{X},t)} \right]^4 + \delta(\vec{X}) \qquad (2.7)$$

where W_u is unfrozen moisture content expressed as a fraction of dry soil weight. The parameters α_u, γ, and δ reflect the character of specific soil-water systems. A nonzero value for δ implies some soil moisture never freezes; for clays and fine silts this is a good approximation down to moderately low temperatures (order 0°F). The form of Equation 2.7 has been tested satisfactorily by fitting it to experimental data. Its form also results in simple expressions for the variational integrals developed in the next section.

The latent energy content used in Equation 2.1 is related to unfrozen moisture content by

$$s(\vec{X},T) = \begin{cases} h_f\ (\vec{X}) & ;\ T(\vec{X},t) > T_f \\ h_f\ (\vec{X})\ W_u(\vec{X},T)/W_t(\vec{X}); & T(\vec{X},t) < T_f \end{cases} \qquad (2.8)$$

where T_f is the nominal freezing point, h_f is the latent energy per unit volume associated with complete thawing of moisture in the soil, and W_t is total moisture content expressed as a fraction of dry soil weight.

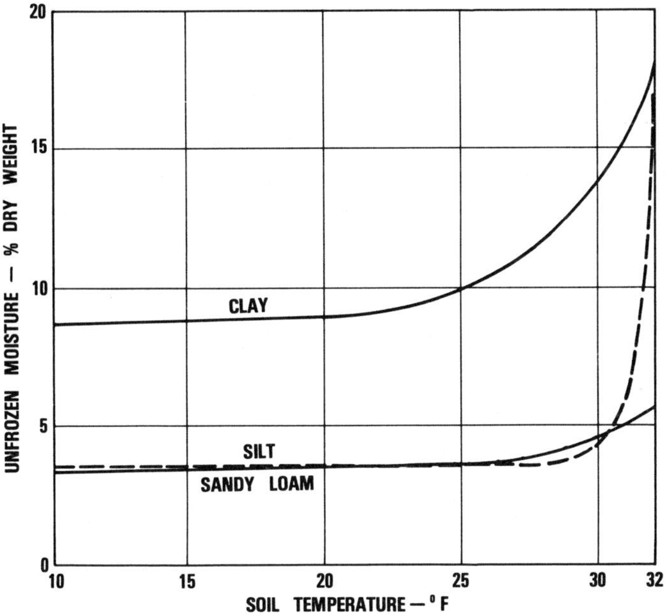

FIG. 1. UNFROZEN MOISTURE CONTENT OF SOME TYPICAL SOILS.

Thermal processes occurring at the interface between the soil surface and the atmosphere are crucial to the analysis of heat transfer in permafrost. The very existence of permafrost depends on the balance that is reached among these processes and with the geothermal flux of energy from the earth's interior. The more important processes are air convection (both natural and forced), conduction through snow, the melting or sublimation of snow, the evaporation of surface and ground water, both incoming shortwave radiation and outgoing longwave radiation, conduction and convection through mats of vegetation, and energy convection by falling snow or rain. Even if these processes were well understood, quantitative analysis would require specific data that are rarely available.

Three methods of approximating heat transfer at the soil surface are available:

1. Estimate the soil surface temperature as a function of time and explicitly impose this temperature by application of Equation 2.2. Measurements of soil surface temperatures in the field both under naturally occurring conditions and near man-made disturbances of the thermal regime provide a basis for the estimate. The main objections to this approach are that to some extent the temperature distribution sought is imposed rather than calculated and that climatic conditions such as snow thickness cannot be readily varied.
2. Estimate an "equivalent air temperature" and a heat transfer coefficient that couples the equivalent air temperature to the soil surface temperature. Thus Equation 2.2 is applied with

$$q_b = h_e (T_e - T) \tag{2.9}$$

where h_e and T_e may be functions of time and location on σ_b. Here, the soil surface temperature T is determined implicitly and some of the parameters that characterize the surface may be varied.
3. Model the individual surface processes as precisely as the state of the art will allow and solve the resulting nonlinear equations simultaneously with the soil conduction equation. This is theoretically the best alternative but now q_b is a complicated relationship describing up to a dozen thermal processes acting in series and parallel. In practice, considerable effort is required to construct the model and to acquire descriptive parameters of sufficient accuracy to justify its use.

Extensive use was made of each of these methods in the design of the Trans-Alaska Pipeline but the details are beyond the scope of the present paper. Scott [10] and Wheeler [12] have described applications of the equivalent air temperature method. Miller [7] has described method 3 as it was applied to the pipeline.

3. VARIATIONAL SOLUTION

The principal advantage of the variational technique for the solution of Stefan's problem is its capability for tracking a melt front within a multidimensional grid element. This capability is inherent since the technique defines a continuous temperature distribution and the melt front is, by definition, the locus of the 32°F isotherm. In contrast, finite difference methods define temperature at only discrete locations and thus inherently determine only the grid elements in which a melt front exists - and not the location of the front within these elements. Interpolation formulas can be devised to locate a melt front within a finite difference element [6], but these formulas are quite complex and it is not apparent that an implicit formulation is practical. A second advantage of the variational technique is its versatility with respect to geometry.

Use of the variational technique to solve Stefan's problem is not entirely novel. Biot and Daughaday [2] used a variational technique to solve the problem of a semi-infinite medium subjected to a constant rate of heat input at the melting surface. They approximated temperature in the solid by a cubic with two unknown parameters and did not address the problem of approximating an arbitrary temperature distribution in two or three dimensions. Hwang, Murray, and Brooker [5] constructed a two-dimensional finite element model for the solution of permafrost thermal problems. Their model is very similar to the model described in this paper but differs in the treatment of Dirichlet boundary conditions and time discretization.

A variational technique for solving the equations defined in the preceding section can be defined as follows: The true solution T is approximated by

$$T(\vec{X},t) \approx u(\vec{X},t) = \sum_{m \,\varepsilon\, M} C_m(t)\ \omega_m(\vec{X}) \ ; \ \vec{X} \,\varepsilon\, R \cup \sigma \tag{3.1}$$

where the "basis functions" ω_m are known functions of position and the coefficients C_m are unknown functions of time. The notation $\sum_{m \,\varepsilon\, M}$ implies summation over all indices m contained in the set M. The set $\{\omega_m\}_{m \,\varepsilon\, M}$ must be linearly independent for $\vec{X} \,\varepsilon\, R$. The choice of boundary conditions is restricted by the "essential condition" that

$$u(\vec{X},t) = \sum_{m \,\varepsilon\, M_a} C_m\ (t)\ \ \omega_m\ (\vec{X}) = T_a(\vec{X},t) \ ; \ \vec{X} \,\varepsilon\, \sigma_a \tag{3.2}$$

where $M_a \subset M$ and the set $\{\omega_m\}_{m \,\varepsilon\, M_a}$ is linearly independent for $\vec{X} \,\varepsilon\, \sigma_a$. The energy flux through the boundary is approximated by

$$\vec{\eta} \cdot \vec{q} \approx \Gamma\ (\vec{X},t,u) = \begin{cases} q_b\ (\vec{X},t,u) & ; \ \vec{X} \,\varepsilon\, \sigma_b \\ \sum_{m \,\varepsilon\, M_a} \gamma_m(t)\ \omega_m(\vec{X}); & \vec{X} \,\varepsilon\, \sigma_a \end{cases} \tag{3.3}$$

The Galerkin equations

$$\int_R L(u)\ \omega_n\ d\ V - \int_\sigma [\Gamma + \vec{\eta} \cdot K\ \vec{\nabla}\ u]\ \omega_n\ d\ A = 0 \ ; \ n \,\varepsilon\, M \tag{3.4}$$

together with Equation 3.2 then determine the unknown coefficients C_m and γ_m. Equation 3.4 can be written in the somewhat more convenient form

$$\int_R \left\{\frac{\partial}{\partial t}\ (\alpha u + s)\ \omega_n + K\ \vec{\nabla}\ u \cdot \vec{\nabla}\ \omega_n\right\} d\ V = \int_\sigma \Gamma\ \omega_n\ d\ A; n \,\varepsilon\, M \tag{3.5}$$

by using Equation 2.1 and integrating by parts.

The backward difference technique is applied to discretize Equation 3.5; the result is

$$\int_R \left\{[(\alpha u+s)^N - (\alpha u+s)^{N-1}]\ \omega_n + \Delta t^N (K\vec{\nabla}u)^N \cdot \vec{\nabla}\ \omega_n\right\} dV$$
$$= \Delta t^N \int_\sigma \Gamma^N\ \omega_n\ d\ A \ ; \ n \,\varepsilon\, M,\ N > 0 \tag{3.6}$$

where a superscript N on a quantity indicates the quantity is evaluated at time level N. Equation 3.6 represents a system of nonlinear algebraic equations in the unknown coefficients C_m^N and γ_m^N which are implicit in u^N and Γ^N according to Equation 3.1 and 3.3. The system can be solved by Newtonian iteration as follows: Let a superscript k on a quantity denote the kth approximation to the value of the quantity at time level N with k equal to zero implying evaluation at time N-1. Define a set of "residuals" by

$$R_n^{k-1} = - \int_R \left\{ \left[(\alpha u+s)^{k-1} - (\alpha u+s)^{N-1} \right] \omega_n + \Delta t^N (K\vec{\nabla})^{k-1} \cdot \vec{\nabla} \omega_n \right\} dV$$

$$+ \Delta t^N \int_{\sigma_b} \Gamma^{k-1} \omega_n \, dA \; ; \; n \in M, \; k>0 \qquad (3.7)$$

The partial derivatives

$$A_{m,n}^{k-1} = - \left(\frac{\partial R_n}{\partial C_m} \right)^{k-1} \; ; \; n, m \in M - M_a \; , \; k > 0 \qquad (3.8)$$

form the elements of a coefficient matrix. The elements of a "correction" vector are defined by

$$E_m^k = C_m^k - C_m^{k-1} \; ; \; m \in M - M_a \; , \; k > 0 \qquad (3.9)$$

The system of equations

$$\sum_{m \in M - M_a} A_{m,n}^{k-1} E_m^k = R_n^{k-1} \; ; \; n \in M - M_a \qquad (3.10)$$

is then linear in the unknowns, E_m^k, and can be solved by point successive overrelaxation or a direct method. Improved values of C_m^N are obtained from

$$C_m^k = C_m^{k-1} + E_m^k * A_f \; ; \; n \in M - M_a \qquad (3.11)$$

where A_f is an attenuation factor chosen by the program so as to limit melt front movement in any one iteration and thereby improve the radius of convergence of the Newtonian iteration. The set $\{C_m^N\}_{m\varepsilon M_a}$ is determined uniquely by Equation 3.2, i.e. by specified boundary temperatures. The iteration process is repeated until

$$\max_{n\varepsilon M-M_a} \left(R_n^{k-1} \;/\; A_{n,n}^{k-1} \right) < CV \tag{3.12}$$

where CV is a constant (typically CV = .0005°F).

Normally the local distribution of the energy flux through σ_a into R is not required; i.e. individual values for γ_m need not be evaluated. The total flux Q_a through σ_a is often needed and can be determined without evaluating the set $\{\gamma_m\}_{m\varepsilon M_a}$. This requires only that the choice of basis functions satisty

$$\sum_{m \,\varepsilon\, M_a} \omega_m(\vec{X}) = 1 \;;\; \vec{X} \,\varepsilon\, \sigma_a \tag{3.13}$$

since then

$$Q_a^N = \int_{\sigma_a} \Gamma^N \, dA = \sum_{m\varepsilon M_a} \int_{\sigma_a} \Gamma^N \omega_m \, dA = - \sum_{m \,\varepsilon\, M_a} R_m^N \tag{3.14}$$

where R_m^N is the final value of R_m^{k-1} for time step N. The last step in Equation 3.14 is accomplished by subtracting Equation 3.7 from Equation 3.6.

The particular basis functions ω_m used in the simulator are piecewise linear functions of two of the three coordinates that determine $\vec{X}$. Thus the simulator is two dimensional even though the equations above appear to apply equally well to three dimensions. One basis function is centered at each node of a rectangular grid spanning R. Each element of the rectangular grid is divided into two triangles by a diagonal. Either 4 diagonals or none pass through a node. The nonzero domain of a particular basis function is the union of triangles having vertices at the node on which basis function is centered.

The basis functions are continuous and satisfy Equation 3.13 for $\vec{X} \in R \cup \sigma$. Additional details are available in Reference [12].

The simulator was coded to facilitate the use of any orthogonal curvilinear coordinate system but, to date, only cartesian and cylindrical coordinates have actually been used. For these two systems the integrals above can be evaluated analytically.

The validity of the numerical procedure outlined above was established by comparing results with analytical solutions [12], laboratory tests, field tests [8], and results produced by independent numerical procedures. No attempt has been made to theoretically justify the procedure as a whole. The use of Equation 3.3 on Dirichlet boundaries is somewhat novel; Douglas, Dupont, and Wheeler [3] have demonstrated theoretically that approximation of flux by a linear combination of basis functions is significantly more accurate than obtaining it from the gradient of u.

4. THERMAL VERTICAL SUPPORT MEMBERS

About 420 miles of the Trans-Alaska pipeline is elevated above ground; along roughly 330 miles of the elevated line, heat pipes were installed in the pile supports to insure an adequate foundation. These thermally protected supports are commonly referred to as "thermal vertical support members" or simply TVSM.

Experience in permafrost areas has shown that foundation problems are often encountered even in the absence of an active heat source in the ground. Permafrost melting may occur due to construction disturbances of the natural surface (e.g. removal of vegetation or installation of a gravel work pad) or from natural causes (e.g., climate changes or forest fires). These problems are common in regions of discontinuous permafrost such as central Alaska because permafrost where it does exist tends to have a temperature near 32°F.

Frost jacking is also a problem in cold climates. Where the soil is initially thawed, several feet of soil near the surface freezes each winter. Ice lens formed during freezing can cause soil to heave around a pile support and cause it to move upward. The upward movement may amount to only an inch or so each year but it is cumulative and ultimately can damage the structure.

Alyeska installed two heat pipes in each of more than 61,000 vertical supports to counter permafrost thaw and frost jacking. A schematic of one of these TVSM is shown in Figure 2. The heat pipes are simply sealed tubes containing ammonia liquid and vapor. Heat entering the lower warmer end of the tube vaporizes liquid; the vapor flows to the upper cooler end and condenses; the condensate then returns down the wall of the tube due to gravity. A radiator at the upper end dissipates latent heat of condensation to the atmosphere. During the winter the devices efficiently transfer heat from soil around the support to the atmosphere. During the summer the devices are inactive since there is no mechanism for moving liquid from the cool lower end of the tube to the warm upper end. Sufficient heat is extracted from the soil each winter to prevent permafrost thaw near a support through the following summer. If the soil is initially thawed, a vertical cylindrical freeze front rather than a horizontal front forms around a support during winter so frost jacking also is prevented.

The permafrost temperature near a TVSM rises to within a few degrees of 32°F each summer. Moreover, the soil temperature at the support at the end of summer is a strong function of the unfrozen moisture characteristics of the soil. Figure 3 indicates that the average temperature at a support and below the active layer may be 1.5°F colder in a clay than in a sand under identical calculations conditions.

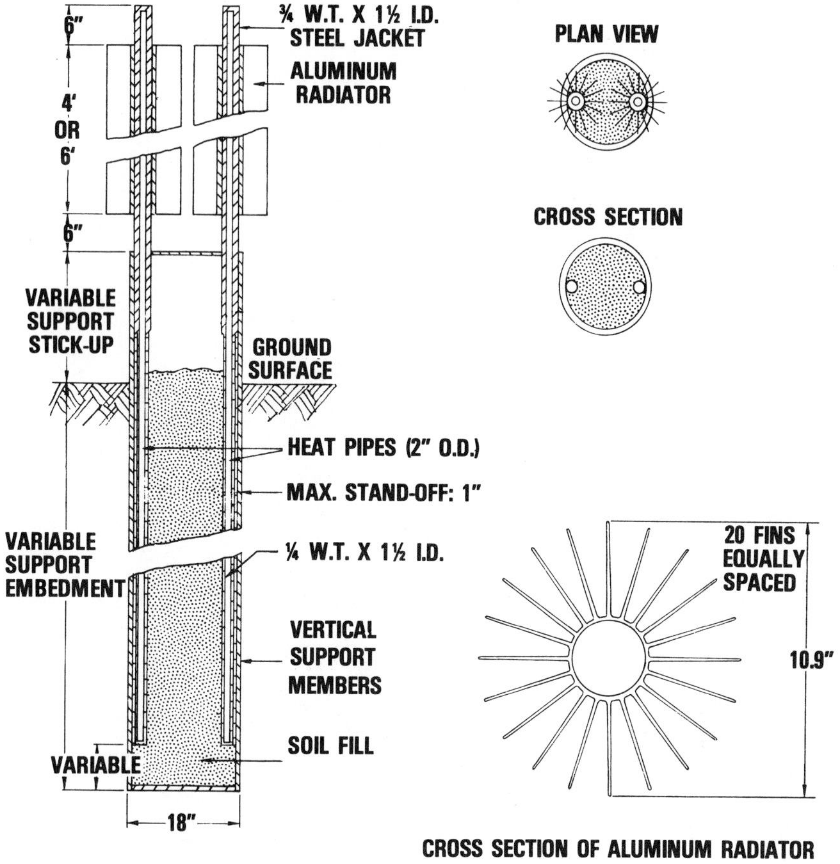

FIG. 2. THERMAL VERTICAL SUPPORT MEMBER.

This sensitivity to soil type plus a tendency for frozen soils to creep at temperatures very near 32°F and the new nature of the technology caused Alyeska to proceed cautiously in applying the TVSM concept to the pipeline. Extensive thermal simulations were carried out to predict the conditions under which TVSM would perform with an adequate safety margin. Laboratory tests were made to evaluate heat pipe-radiator combinations and other devices [4]. Six manufacturers of thermal devices conducted tests to establish that their products would perform satisfactorily over the thirty-year design life of the pipeline [11]. Finally, a series of three field tests were conducted in Alaska to verify TVSM performance under field conditions and validate design calculations [8].

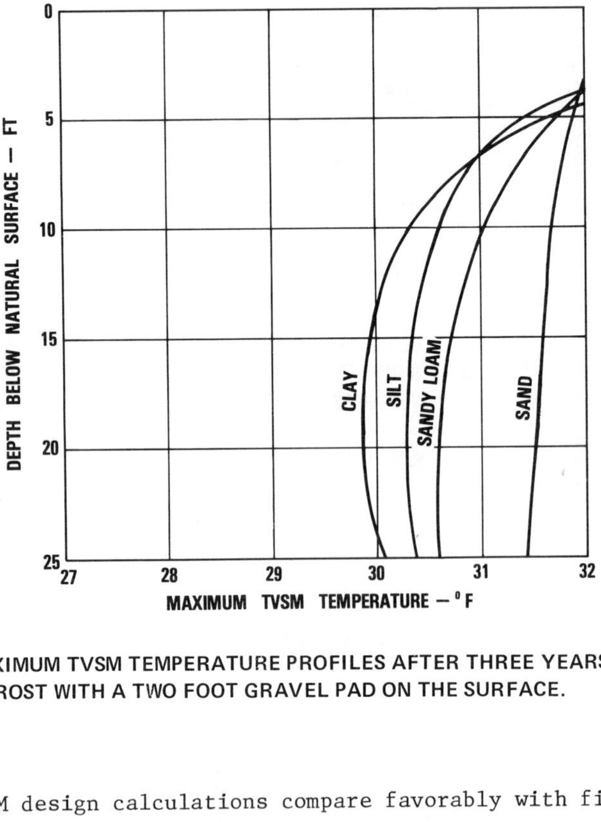

FIG. 3. MAXIMUM TVSM TEMPERATURE PROFILES AFTER THREE YEARS FOR 31°F PERMAFROST WITH A TWO FOOT GRAVEL PAD ON THE SURFACE.

TVSM design calculations compare favorably with field observations. Figure 4 compares design temperature at a depth of 20 ft with temperatures measured on two TVSM at a site near Fairbanks. After one winter of heat pipe operation, the maximum measured temperature was 29.54°F; the corresponding calculated temperature was 30.18°F. Heat pipe performance and weather conditions were conservatively modeled so it is to be expected that higher temperatures were predicted than were observed.

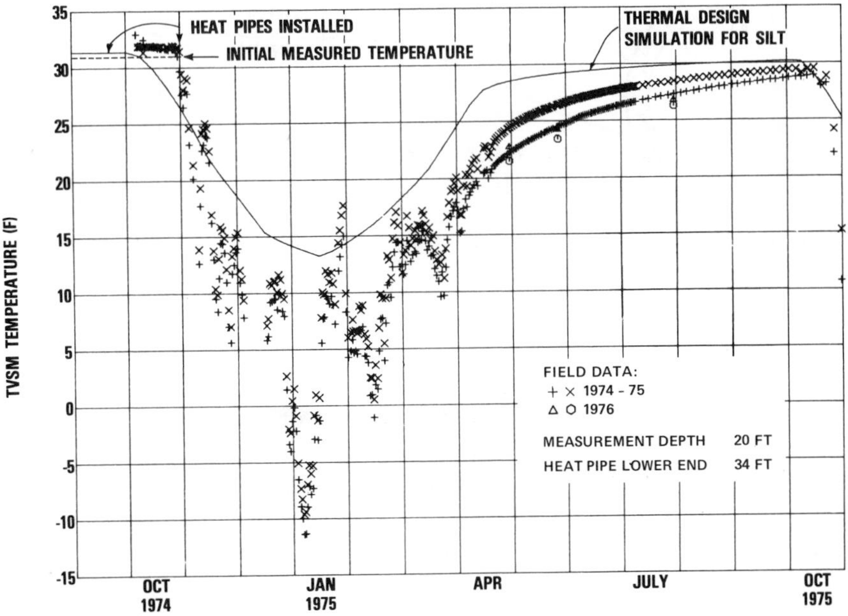

FIG. 4. COMPARISON OF TVSM DESIGN CALCULATIONS WITH FIELD DATA.

5. THAW PREVENTION BY MECHANICAL REFRIGERATION

Burial of an insulated line traced with refrigeration lines was an alternative to elevated line early in the Alyeska project. Ultimately, economics favored elevated line and burial with mechanical refrigeration was used on only four miles of line where road and animal crossings made elevation impractical.

Refrigeration lines below an insulated pipeline can absorb energy that escapes through the insulation and thereby prevent permafrost thaw below the pipeline. A calculated thaw front location in October after five years of operation is shown in Figure 5 for a typical case. Maximum depth of the front has

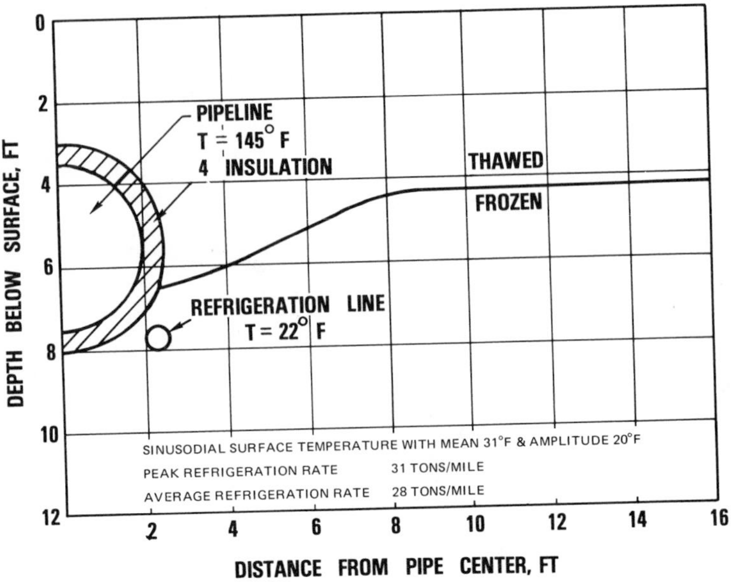

FIG. 5. THAW BULB FOR SOIL REFRIGERATION SYSTEM IN WARM (31° F) PERMAFROST.

stabilized but it does dip down near the pipe. Additional refrigeration lines can be added at bends for better lateral support. Figure 5 shows four inches of insulation on the pipeline but three inches was found to be more economical even though more refrigeration was required. Less than three inches of insulation does not allow adequate repair time if the mechanical equipment fails. Wheeler [12] has presented a parameter study of factors influencing thaw prevention by mechanical refrigeration.

6. CONCLUSIONS

1. Thawing of ice-rich permafrost below a pipeline or other facility may result in structural damage and thus should be considered in design.

2. The variational method is well suited for the solution of permafrost heat transfer problems.
3. Extensive thermal analysis was required to insure a sound foundation for the Trans-Alaska pipeline in areas of ice-rich permafrost.

REFERENCES

1. D. M. Anderson and N. R. Morgenstern. "Physics, Chemistry and Mechanics of Frozen Ground," North American Contribution Permafrost Second International Conference, National Academy of Sciences, Washington, DC, (1973).

2. M. A. Biot and H. Daughaday. "Variational Analysis of Ablation," J. Aerospace Sci., v. 29, 1962, pp. 227-228.

3. J. Douglas, Jr., T. Dupont, M. F. Wheeler. "A Galerken Procedure for Approximating the Flux on the Boundary for Elliptic and Parabolic Boundary Value Problems", Revue Francaise d' Automtique Informatique et Recherche Operationnelle, v. 8, 1974, pp. 61-66.

4. J. W. Galate. "Passive Refrigeration for Arctic Pile Supports", ASME paper 75-PET-26 presented 13th Pet. Mech. Engr. Conf., 1975.

5. C. T. Hwang, D. W. Murray, and E. W. Brooker. "A Thermal Analysis for Structures on Permafrost", Canadian Geotechnical J., v. 9, 1972, pp. 33-46.

6. A. Lazaridis. "A Numerical Solution of the Multidimensional Solidification (or Melting) Problem", Int. J. Heat Mass Transfer, v. 13, 1970, pp. 1459-1477.

7. T. W. Miller. "The Surface Heat Balance in Simulations of Permafrost Behavior", ASME paper 75-WA/HT-86 presented Winter Annual Meeting, 1975.

8. S. W. Pearson. "Thermal Performance Verfication of Thermal Vertical Support Members for the Trans-Alaska Pipeline", ASME paper 77-WA/HT-34 to be presented Winter Annual Meeting, 1977.

9. T. L. Péwé. "Permafrost: Challenge of the Arctic", _1976 Yearbook of Science and the Future_, Encyclopedia Britannica (1975).

10. R. F. Scott. Predicted Depth of Freeze or Thaw by _Climatological Analysis of Cumulative Heat Flow_, U. S. Army Cold Regions Research and Engineering Laboratory Technical Report 195, Hanover (1969).

11. E. D. Waters, C. L. Johnson, and J. A. Wheeler. "The Application of Heat Pipes to the Trans-Alaska Pipeline", paper WD2535 presented 10th Energy Conversion and Engineering Conf., 1975.

12. J. A. Wheeler. "Simulation of Heat Transfer From a Warm Pipeline Buried in Permafrost", AICHE paper 27b presented 74th National Meeting, 1973.

J. A. Wheeler
Exxon Production Research Company
Post Office Box 2189
Houston, Texas 77001

MOVING BOUNDARY PROBLEMS

HEAT TRANSFER CHARACTERISTICS IN A MODEL ICE-WATER HEAT SINK

Yin-Chao Yen

An experimental study of forced convective heat transfer over a vertical melting plate 0.914 m high and 0.305 m wide has been conducted. This study covers the water velocity ranging from 1.7 to 9.8 mm/s and bulk water temperature from 1.11^o to 7.5^oC. The test column was constructed in such a way that it could be considered to be a segmental ring cut from the annual ring of the prototype heat sink. This was accomplished by installing two side-moving belts adjusted to move in the same direction and at the same velocity of the water. The experimental results are correlated in terms of Nusselt, Prandtl and Reynolds numbers and can be expressed by $Nu_L/Pr = 3.275(Re_L)^{0.270}$ with a moderate correlation coefficient of 0.843. The results are expected to be useful in predicting the heat transfer characteristics of the much larger prototype ice-water heat sink.

1. INTRODUCTION

Ice has been considered a promising heat dissipating medium for enclosed power generating systems because of its large heat of fusion associated with phase change. In particular, considerable interest has been developed in investigating the feasibility of using ice as a heat sink for underground hardened military installations. These sinks would be used in the event of a bottom-up (self-contained) situation to absorb the waste heat. During the plant's normal operation, the heat rejection process is commonly done at the ground surface with cooling towers and water reservoirs. Unlike the normal mode, this stand-by system must be ready and put into use instantaneously and be capable of absorbing all the waste heat during a certain short period of time. One proposed heat rejection system for a typical underground installation consists of a steam condenser and three ice-water heat sinks (connected either in a series or parallel). The exhaust steam from the turbine condenses and rejects the heat to the water flowing

ISBN 0-12-757350-X

through the condenser. This cooling water would then dissipate the heat to the sinks by both melting the ice and heating the water. Relatively cool water would then be drawn from the heat sinks and pass through the condenser. One concept in particular is using cylindrical ice columns of 33.5 m in height and 19.8 in diameter, with a circulating coolant rate of 9.08 x 10^5 Kg/hr and a minimum anticipated heat rejection rate of 6.73 x 10^6 Kcal/hr. It is important to notice the interdependent nature of the heat sink and the power generating system's performance. The power system's thermal efficiency is influenced by the coolant water temperature; in general, the lower the sink temperature, the more efficient the power system. Also the total heat rejection rate to the sink is evidently a function of the power system's thermal efficiency.

Brown and Quinn [2] studied this problem by scaling their experimental setup to the actual prototype with a cylindrical block of ice 1.75 m in height and 1.22 m in diameter. Since the same fluid is used for both the model and the prototype, it is impossible to devise a model with dynamic similarity to the prototype. To circumvent this difficulty, they developed a scale model taking into consideration the temperature, heat transfer and time factors, in addition to the usual geometrical scaling, as a basis of analysis for the prototype. The two essential scaling relationships are $(q_r)_p=(R_oL_o)_p/(R_oL_o)_m(q_r)_m$ and $(M)_p=(R_oL_o)_p/(R_oL_o)_m(M)_m$ where $(\ )_p$ and $(\ )_m$ refer to the prototype and the model respectively. q_r and M are the heat rejection and coolant flow rate, and R_o and L_o are the original radius and the height of the cylinder. The results of the experimental model studies were compared with those from the numerical analysis of the same problem. Favorable comparison was obtained providing the confidence in using model study results to predict the prototype performance. Stubstad and Quinn [8] investigated this problem further by conducting a larger scale experiment using an ice cylinder 3.0 m in height and 1.83 m in diameter. Both studies have provided information on melting patterns and the characteristics of the coolant water temperature as a function of operating time.

In a more recent study, Grande [4] presented an analysis and development of the conceptual design of an ice-water heat sink system to accommodate the closed cycle operation of a 1500 KW nuclear power plant in a hardened underground installation. This study also provides a critical review of previous work on this subject. It also extends the analytical description of the sink's thermal performance and assesses the available options for varying the flow path of coolant water within the sink. Computer programs were developed for predicting the time histories of heat sinks and a practical sink design was developed based on power plant performance constraints.

Most of the previous work involving melting of ice was on free convective heat transfer. This work is complicated by water's intriguing feature of possessing a maximum density at about $4^{o}C$. Since the existence of maximum density, in melting or freezing cases, implies that the thermal expansion coefficient β changes its sign, the use of an average value of β thus becomes physically unrealistic. More important is the possibility of dual flow due to the change in direction of the buoyancy force. Another factor is phase change at the solid-liquid interface, which introduces an interfacial velocity and may distort the flow field, thereby affecting the heat transfer rate. Tkachev [9] was the first to observe a minimum Nusselt number at $5.5^{o}C$ in his experimental work with melting cylinders. He attributed this minimum to the possibility of dual flow within the boundary layer. Schechter and Isbin [6] experimentally demonstrated the existence of dual flow. From their measurements of the velocity profiles near a warm plate immersed in water, they showed that at certain temperatures the inner portion of the boundary layer moved upwards while the outer portion moved downwards. Merk [5],considering water as a high Prandtl number fluid and using a moving coordinate system, solved the problem by using an integral method. For the case of ice spheres, Merk's analysis indicated that there existed a minimum Nusselt number and a complete inversion of flow at a bulk temperature of 4.8^{o}C.

The present study was aimed at simulating the exact flow conditions anticipated within the prototype. An inherent problem with scaling a flow situation such as this, is that the flow dynamics of a scale model cannot be exactly reproduced in the prototype. To circumvent this difficulty, the present experimental setup is designed in such a manner that it is essentially a segment cut out of the prototype. This is accomplished by considering the ice plate as the vertical ice surface of the column, and the back wall of the test tank as the containment wall of the prototype. The fact that within the annular space the experimental segment is bounded on both sides by water flowing down at the same rate is taken into account with the installation of continous belts on opposing sides of the test tank designed to move at the same velocity of the water.

2. EXPERIMENTAL APPARATUS

A General Considerations

To minimize heat loss in pumping water through the flowmeter to the inlet of the test assembly, PVC piping was used throughout for its low thermal conductivity. The high corrosion resistance of aluminum, stainless steel, and brass made their use practical in various load bearing frames and components that were in contact with the water. A small piece of magnesium was placed in the reservoir as a sacrificial metal to protect the aluminum from cathodic reaction with the brass and stainless steel. Distilled water was used to eliminate the problem of chlorine corrosion found in tap water and the production of metal salts that could influence the ice melting rate. Teflon and nylon were used for the bearing surfaces, and close sliding fits between assembly parts were designed to eliminate the need for lubricants which could contaminate the water and affect its physical properties. Apparatus components containing oil, such as gear reducers and flexible shafts were isolated from the rest of the system to preclude the possibility of alternating the viscosity and surface tension of the circulating water, possibly adding unknown factors into the ice melting characteristics. Figure 1 shows the schematic of the experimental setup. It essentially consists of four major sections, i.e. the water reservoir, test chamber assembly, and pumping and control systems.

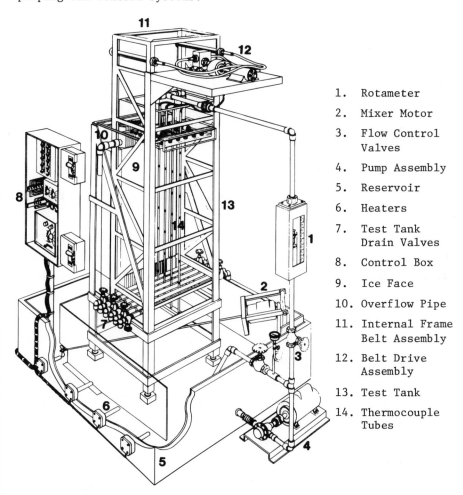

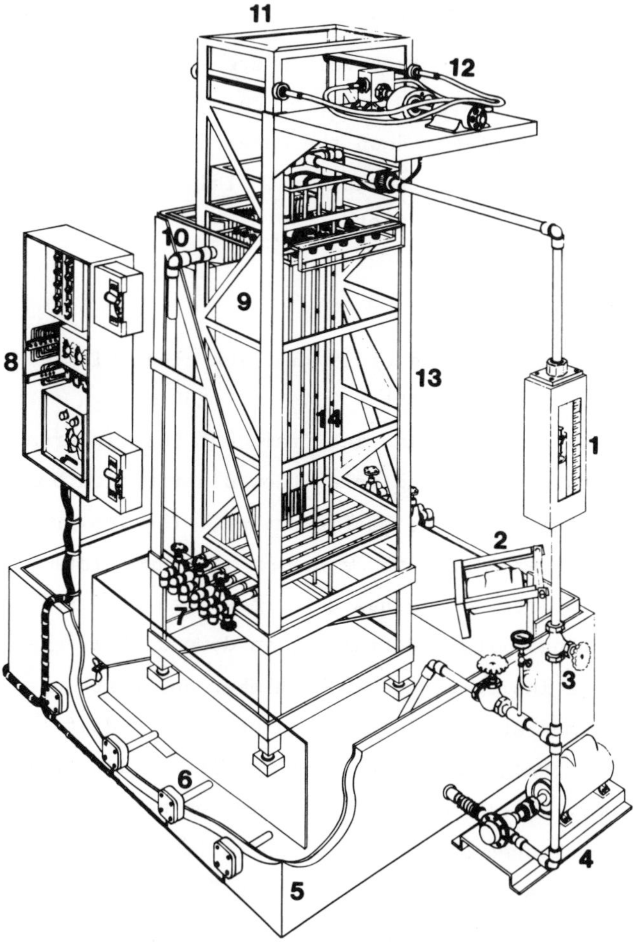

1. Rotameter
2. Mixer Motor
3. Flow Control Valves
4. Pump Assembly
5. Reservoir
6. Heaters
7. Test Tank Drain Valves
8. Control Box
9. Ice Face
10. Overflow Pipe
11. Internal Frame Belt Assembly
12. Belt Drive Assembly
13. Test Tank
14. Thermocouple Tubes

Figure 1. Experimental Apparatus

B Water Reservoir

The 1.22 x 1.22 x 0.46 m high tank is constructed of welded 4.76 mm aluminum plate. It is supported above the floor by a channel aluminum footing. The bottom and sides of the tank are insulated externally by Styrofoam, 1-1/2 in. thick. The tank has an operating capacity of 0.473 m^3 and is equipped with a 1/3 horsepower mixer. The mixer, through the aid of a baffle system installed within the reservoir, drains water past the seven–750 watt heaters on the two adjoining sides of the tank, pulling it through a passage at the end of the intersection of the two baffle plates into the major open section of the reservoir where the water is thoroughly mixed. A thermistor probe, mounted on one of the baffle plates and immersed half-way down the water depth is connected to an on-off controller for the heaters. This probe registers temperature change to within ± 0.05°C.

C Test Chamber Assembly

The test chamber is constructed of 19 mm thick clear Plexiglas with external dimensions of 0.495 x 0.445 x 1.238 m high. It contains the ice frame and ice sample, ice frame baffles, internal frame belt assembly, drain system and inlet flow distribution box. Figure 2 shows the ice frame and the baffle device. The ice frame is constructed of 19 mm Plexiglas with an 8.5 mm aluminum plate mounted in the back. The ice sample has a surface area of 0.914 x 0.305 m exposed to melting. The ice frame slips down within recessed surfaces on the inside of the test chamber and rests upon a baffle arrangement aimed at reducing erosion near the end of the ice sample. A similar arrangement on the top end of the ice frame serves the same purpose. The simulated water flow on both sides of the chamber is accomplished by the use of belts (made of Mylar) moving downward at the same velocity as the water flowing past the ice.

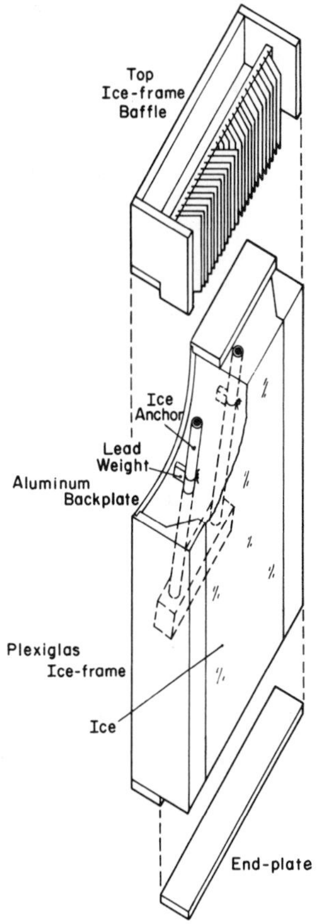

Figure 2. Ice Frame and Baffle Device.

The flow distributor, also constructed of Plexiglas with a perforated bottom, sits on top of the test chamber. Spacers are used to vary the gap between the bottom and the above removable perforated plate. Both plates have drilled holes, each 12.7 mm in diameter and arranged in a mutually staggered fashion. The net effect is to secure an even water flow across the cross section of the test chamber. An even withdrawal of water was accomplished by installing six 19.0 mm ID PVC pipes with slits 0.305 m long and 25.4 mm wide facing the chamber bottom. These pipes are evenly spaced and protrude horizontally out both sides of the test chamber. Twelve PVC gate valves were installed at the ends of the pipes for adjustment of equal flow.

The internal frame belt assembly is placed within the perimeter of the test chamber. The belts, made of 0.089 mm thick Mylar are set up such that there exists exactly a 0.305 x 0.305 m area for water flow. The belts are driven downward at the water velocity to create a condition of an infinite medium of water flowing on both sides of the test chamber. The belt assembly is driven by a 1/12 horsepower variable speed d.c. motor with a speed range of 100 to 1750 rpm. The speed is first reduced by a 10:1 ratio gear reducer with a single takeoff and followed by a

20:1 reducer with double takeoffs which are connected with flexible shafts to the belt rollers of 38 mm diameter. The rollers are designed to vary within the range of 0.5 to 8.75 rpm. The entire test chamber is supported by an external frame centrally situated in the reservoir. The frame is mounted on threaded brass studs on the reservoir bottom and can be adjusted to ensure the whole assembly being vertical. The test chamber is hinged on the external frame so that it can be tilted back for easy removal of the belt drive assembly and the ice frame.

D Pumping and Electrical Control System

A rotary gear pump is connected to the reservoir by a flexible reinforced rubber hose to reduce the intensity of noise and vibration. The desired water rate to the test chamber is regulated with a bypass line and is in conjunction with a Rotameter which is rated from 1.39×10^{-4} to 1.39×10^{-3} m^3/s.

All electrical controls are centralized within a control box. Both 120- and 208-V power is supplied through individual disconnect switches. A variable speed control unit is used to regulate the 1/12 horsepower d.c. motor for the belt drive. A thermistor on-off control connected to the probe operates three relays for the seven heaters. Independent manual control for each of the heaters, the mixer motor, rotary gear pump, as well as the temperature control unit are provided.

3. EXPERIMENTAL PROCEDURE

Initially the ice frame is prepared for the formation of an ice sample. The two end Plexiglas plates are held in place to the ice frame by screws with all the mating surfaces being temporarily sealed with silicone elastomer. The ice anchor (made of two parallel PVC tubings), leading to threaded holes in a rectangular Plexiglas block, is then placed within the ice frame with the tubing ends plugged with rubber stoppers and sealed with silicone elastomer. The anchor is kept from moving out of place by lead weights and the plugged end rests against the end plate. The ice frame is then placed and leveled on the aluminum cold plates; this enables the formed ice surface to be level with the ice frame. In filling the ice frame with distilled water, the volumetric change associated with phase change is also taken into account. Ethylene glycol at $-20^{o}C$ is circulated through the two coldplates in a series from a constant temperature bath cooled by trichloroethylene at $-73^{o}C$. During the freezing period, the ice frame is covered with an inverted, tightly fitting Styrofoam box to ensure the freezing process unidirectional upward, thereby eliminating the possible volumetric expansion problem associated with phase change. Once the ice sample is made, the surface is smoothed and even with a warm copper billet. The ice frame is then placed in a coldroom maintained at $-1^{o}C$ and left there for at least 36 hours with a fan blowing on the backplate to ensure that the ice frame and its ice sample are conditioned to a uniform temperature of $-1^{o}C$.

The end plates and the rubber stopper are then removed from the ice frame and the ice anchor and the aged silicone elastomer are cleared from the ice frame. Two stainless rods are inserted in the anchor tubes and threaded into the Plexiglas block to create a handle for placing the ice frame in and removing it from the test chamber.

At the same time, the drainage valves are calibrated for each specific flow velocity with each one adjusted to equal flow rates at the test temperature. After the calibration, the room temperature is raised to $20^{o}C$ to dry out the water remaining in the valve seats and piping joints to prevent freeze-up when the room is prepared for and maintained at the conditioned ice temperature of $-1^{o}C$ during the experiment. Once the room and the water reservoir have reached the desired temperatures, the ice frame is brought to the room and weighed, the test chamber tilted back, and the ice frame placed into the chamber. The handle is removed and replaced with two rubber stoppers, and the belt assembly is slipped into the chamber. The inlet flow distribution box is attached to the side wall of the chamber and two thermistors are then inserted through the perforated plates and left hanging about one- and two-thirds down the ice frame to measure the water temperature during the experiment. To prevent melting during rapid filling, a thin Plexiglas sheet is inserted between the belt assembly and ice frame, and then the top ice frame baffle is placed right above the ice frame. Finally the chamber is returned to its vertical position and the water inlet header pipe and flexible drive shafts are connected.

The belt drive is turned on while the chamber is being filled, and the experiment commences as soon as the water reaches the overflow, and the Plexiglas sheet is removed. The desired flow rate is regulated by setting the Rotameter. The thermistor readings are recorded periodically to observe the variation of water temperature during the test. The duration of an experiment is primarily determined by the water temperature; for water temperature as low as 1.1°C, it can last as long as six hours while for higher temperatures, the duration is about two or three hours, partially dependent on the water velocity. The experiment is terminated by turning off the gear pump and unplugging the two auxiliary drains at the bottom of the test chamber. At that time the inlet water pipe is disconnected, and the thermistor leads removed. Then the test chamber is tilted back as soon as the water level reaches the chamber bottom, and the inlet flow distributor, the top ice frame baffle and belt drive assembly are subsequently removed. Since, during the test the ice frame and the room are maintained at -1°C, a thin layer of ice may form between the backplate of the ice frame and the chamber. To ease the removal of the ice frame, a small amount of warm water is poured directly at the interface. The ice frame is weighed immediately after its removal from the chamber and the melting surface topography is noted.

4. EXPERIMENTAL RESULTS

Forty-six experiments were conducted covering water velocities v of 1.7 to 9.8 mm/s and bulk water temperature, T_∞ of 1.11° to 7.50°C. The ice surface was found to be relatively flat with some extra melting along the side near the top. This is apparently due to the turbulence created by the inside edges of the ice-frame as ice melts away. For the case of the highest velocity and temperature used in the tests, there is greater melting near the water entrance. A few experiments were run under identical conditions to determine the data reproducibility. Also a few of them were tested without the belts moving to assess the effect of buildup of side boundaries on the melting rate. No significant differences were observed. However, to simulate the prototype operation conditions more closely, all the rest of the experiments were carried out with the belts moving. Figure 3 shows the

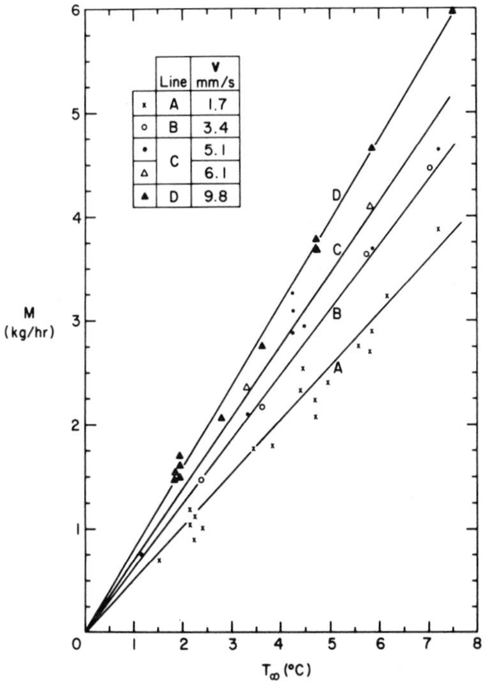

Figure 3. Melting rate as a function of T_∞.

variation of melting rate M with T_∞ using v as the parameter. Lines are drawn among the data and are extrapolated to the origin for no melting at 0^oC. The data show a great deal of scattering, demonstrating the possibility that various extents of turbulence were introduced at the water entrance. To circumvent this, the water must be maintained at an overflow level during the entire test; otherwise, there will be a gap between the distributor and water level and the water will be dropping down to the chamber, thus creating turbulence and promoting a high rate of heat transfer. The low sensitivity of the weighing scale also contributed to the inconsistency of the data. In spite of the scattering, however, the trends clearly indicate that the values of M have a very strong dependence on T_∞ and to a lesser degree on v.

The results are expressed in terms of an average heat transfer coefficient h defined as

$$h = (q_s + q_f)/(A\ \Delta T\ \theta) \tag{4.1}$$

where q_s is the sensible heat needed to raise the ice from initial temperature $T_i(<0^oC)$ to the melting temperature $T_m(=0^oC)$, q_f is the heat of fusion, A is the area of melting surface, ΔT is the temperature difference $T_\infty - T_m$ (since $T_m = 0^oC$, $\Delta T = T_\infty$) and θ is duration of experiment. A Reynolds number $Re_L = (v\rho L/\mu)$ is used to characterize the flow systems where L is the ice height, ρ and μ are the water density and dynamic viscosity, respectively. The physical properties are evaluated at the arithmetic mean of T_∞ and T_m or $T_\infty/2$. The amount of ice melted W is used to compute $q_s = W\ c_i(T_m - T_i)$ and $q_f = WL_f$ where c_i is the specific heat of ice. Figure 4 shows a correlation between Nu_L/Pr and Re_L with the Nusselt and Prandtl numbers defined as $Nu_L = hL/k$ and $Pr = c\mu/k$ in which c and k are the specific heat and thermal conductivity of water. The correlation can be fairly represented by

$$Nu_L/Pr = 3.275(Re_L)^{0.270} \tag{4.2}$$

with a correlation coefficient of 0.843. The relatively low value of correlation coefficient, in addition to reasons mentioned

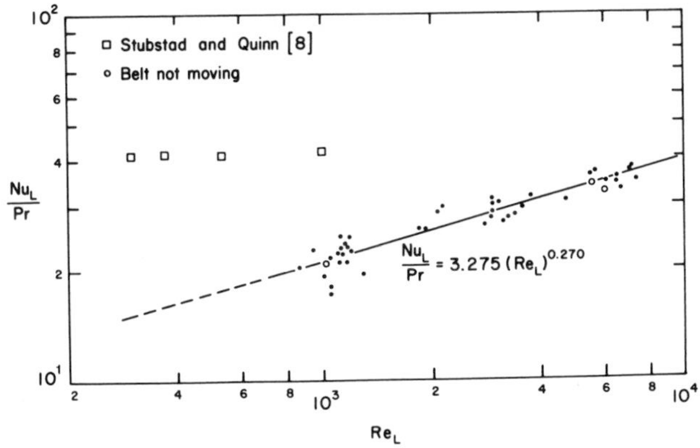

Figure 4. Relationship between Nu_L/Pr and Re_L.

above, may partially result from the difficulty in ascertaining the accuracy in the temperature measurement, especially when T_∞ is close to T_m. The presence of two belts does not affect the heat transfer noticeably and in fact all the data are within the experimental range. A partial explanation is that there would be a minimal (if any at all) frictional resistance to flow and thereby creating a very thin boundary layer. Another explanation is that the ice sample is wide enough to consider the boundary effect insignificant.

5. DISCUSSION AND COMPARISON OF RESULTS

As indicated in the Introduction, there has been no experimental forced convective studies of a vertical melting plate. However, Yen and Tien [11] analytically solved the laminar heat transfer problem over a horizontal melting plate with a linear boundary layer velocity distribution, i.e. $v_x = By$ where v_x is the water velocity in the direction along the plate, B is a constant and y is distance perpenticular to the plate surface. The equation of energy was solved and the value of Nu_L can be expressed by

$$Nu_L = (L/a_n)(B/(9\alpha x)^{1/3} \qquad (5.1)$$

where L is plate length. The value of a_n is evaluated from the integral

$$a_n = \int_0^\infty \exp(-\lambda^3 + \beta\lambda/a_n)d\lambda \tag{5.2}$$

and is found to be a function of β which is defined as $\beta = c(T_\infty - T_m)/L_f$ and can be regarded as a melting parameter, and λ is a dummy variable. By approximating the value of B with τ_w/μ where τ_w is the ice-water interface shear stress, the ratio of Nu_L (with melting) to Nu_0 (without melting) is found to be

$$Nu_L/Nu_0 = (a_0/a_n)(\tau_w/\tau_0)^{1/3} \tag{5.3}$$

where a_0 is defined as

$$a_0 = \int_0^\infty \exp(-\lambda^3)d\lambda \tag{5.4}$$

and τ_0 is wall shear stress without melting. Since the velocity profile is assumed to be comparable to the temperature profile and τ_0 is proportional to the velocity gradient, Eq. (5.3) becomes

$$Nu_L/Nu_0 = (a_0/a_n)^{4/3} \tag{5.5}$$

For the case of laminar flow over a flat plate with no pressure gradient and no mass transfer, Nu_0 is given by Schlichting [7] as

$$Nu_0 = 0.664\ (a_0/a_n)^{4/3}(Pr)^{1/3}(Re_L)^{1/2} \tag{5.7}$$

Figure 5 shows a comparison of Eqs. (5.6) and (5.7) along with the data from the present study. Actual experimental conditions are used to evaluate the value of $(a_0/a_n)4/3$ in Eq. (5.7) which depends on β. Since β varies in the experiments between 0.014 and 0.094, the effect of $(a_0/a_n)4/3$ on Nu_L is therefore rather small. However for $\beta=1$, the reduction in the value of Nu_L amounts to about 42%. As can be seen from the Figure, the present results are approximately four to five times greater than those predicted from a horizontal melting plate. The data can be expressed as

$$Nu_L/(Pr)^{1/3} = 16.5(Re_L)^{0.279} \tag{5.8}$$

with a moderate correlation coefficient of 0.895.

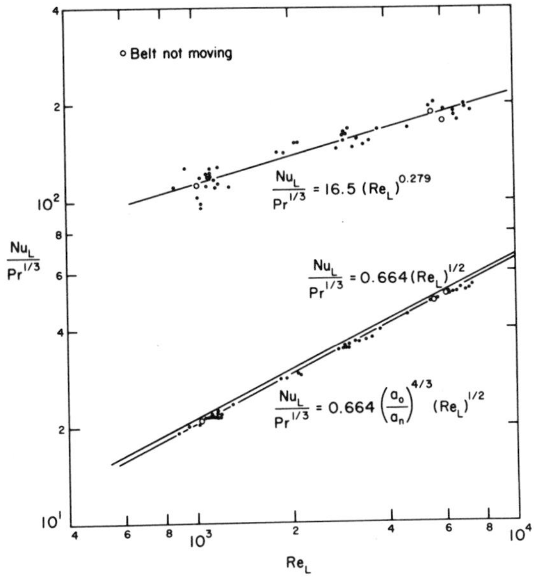

Figure 5. Comparison of Results.

A few data points from Stubstat and Quinn [8] are shown in Figure 5. The disagreement with present data is apparent. In their work, a column of ice 3 m high and 1.83 m initial diameter was used. For a given mass flow rate, the annular flow space increased as melting progressed; therefore, for a given run, Re_L decreased. The computation of h was based on no change in column height, L, and was given by

$$h = (\rho_i L_f / \overline{T}_\infty)(\Delta R / \Delta\theta) \tag{5.9}$$

where ρ_i is the ice density and $\overline{T}_\infty$ is the average bulk water temperature. The incremental values of radius and time $\Delta R = R_{\theta 1} - R_{\theta 2}$ and $\Delta\theta = \theta_2 - \theta_1$ are related by

$$\Delta R = R_{\theta 1} - \{(R_{\theta 1})^2 - (\Delta\theta/\rho_i \pi L)(dM_i/d\theta)\}^{1/2} \tag{5.10}$$

where M_i is the initial ice mass in the tank. The discrepancy between these and the present data can be attributed largely to the assumption of constant column height and the accuracy of the measured water level variations accompanied with phase change to compute the variation of R. The higher values of Nu_L can also be partially attributed to the high degree of turbulence created by the entering water which was sprayed directly down from the nozzle into the annular space.

As mentioned earlier, a number of researchers have worked on the free convection melting heat transfer. Vanier and Tien [10] studied the problem analytically and experimentally. By various combinations of plate temperature T_p and T_∞, they were able to identify the various flow regions created by the exhibition of a density maximum at about 4^oC. By using boundary layer approximations, in conjunction with similarity transformation, they solved the equations of motion and energy numerically and the Nu_L is found

$$Nu_L = 2(2)^{1/2}/3(Gr)^{1/4}[-H'(0)] \tag{5.11}$$

where Gr is Grashof number and is always positive and is defined as $Gr=gL^3|\beta_\infty(T_p-T_\infty)|/\nu^2$, $H'(0)$ is the dimensionless temperature gradient at the plate, and ν is the kinematic viscosity. β_∞ is defined by

$$\beta_\infty=(\beta_1+2\ \beta_2T_\infty + 3\beta_3T_\infty^2)/(1+\beta_1T_\infty+\beta_2T_\infty^2 + \beta_3T_\infty^3) \tag{5.12}$$

in which β_1, β_2, and β_3 are the coefficients in the cubic expression of the density-temperature relationship of water. For the case $T_p = 0^oC$ (equivalent to the case of melting $T_m=0^oC$), Vanier and Tien's numerical results (after scaling to the present experimental plate height) are shown in Figure 6.

The most recent experimental study of free convective heat transfer over a vertical ice slab immersed in cold water has been reported by Bendell and Gebhart [1]. In this study, the effect of T_∞, especially in the vicinity of the density maximum, on heat transfer was determined. Also the effect of the density maximum on flow direction and T_∞ where the flow changes direction was determined. Their results were compared with the analytical work of Gebhart and Mollendorf [3] and only a mean magnitude of

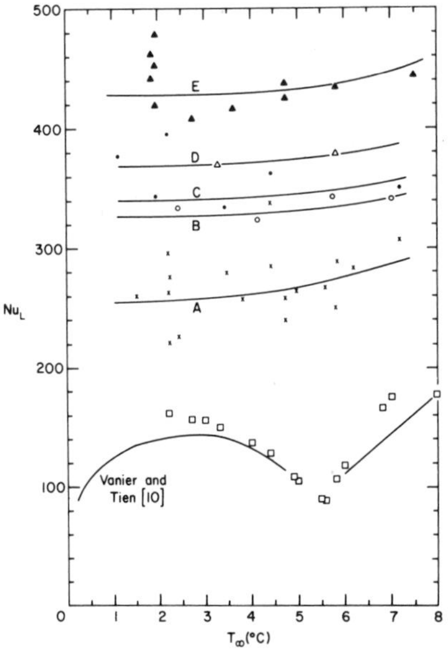

Figure 6. Nu_L as a function of T_∞. x, curve A, v=1.7mm/s; o, curve B, v=3.4 mm/s; ●, curve C, v=5.1 mm/s; Δ, curve D, v=6.1 mm/s; ▲, curve E, v=9.8 mm/s; □, data of Bendell and Gebhart [1].

deviation (5.6%) was found. Their results are also shown in Figure 6 along with the present results. It can be seen that Bendell and Gebhart's results agree fairly well with those of Vanier and Tien. It can be clearly noted that, for the case of forced motion, there is a significant increase in heat transfer over free convection even with a minimal velocity of 1.7 mm/s. The greater scattering of data, especially at the lower end of the temperatures covered in the experiment, can be partially attributed to the inaccuracy in its temperature measurement. However, the effect of flow on the melting heat transfer can easily be visualized in Figures 3, 4 and 5. It also can be concluded that the change in flow direction due to buoyancy forces has exhibited no significant effect on heat transfer, as indicated numerically by Vanier and Tien and experimentally by Bendell and Gebhart.

6. CONCLUSIONS

1. An experimental device to simulate the rather large dimensions of a prototype heat sink has been developed. This was accomplished by installing belts on two opposite sides of a square column. The belts were designed and adjusted to move in the same direction and at the same velocity as the water. In doing this, the laboratory setup can be considered as a segmental ring cut out from the proposed prototype.

2. For bulk water temperature ranging from 1.11^{o} to $7.5^{o}C$ and water velocity from 1.7 to 9.8 mm/s, the present result can be fairly represented by

$$Nu_L/Pr = 3.275(Re_L)^{0.270}$$

with a moderate correlation coefficient of 0.843. It is expected that this expression will predict the heat transfer characteristics of the prototype even though the data were taken from experiments with one specific plate height.

3. It also can be concluded that, in the case of forced convective melting heat transfer, the effect of buoyancy forces caused by the density maximum at $4^{o}C$ appeared insignificant. This is evident even at a very low velocity, as the heat transfer is drastically reduced in the vicinity of $5.6^{o}C$ in the case of free convective melting heat transfer.

REFERENCES

1. M.S. Bendell and B. Gebhart. Int. J. Heat Mass Transfer, (1976), pp. 1081-1087.

2. J.L. Brown and W.F. Quinn. An Annular Flow of Ice-Water Model Heat Sink, U.S. Army Cold Regions Research and Engineering Laboratory (USACRREL) Special Report 236, Hanover, NH, 1975.

3. B. Gebhart and J. Mollendorf. Submitted to J. Fluid Mech. (1976) cited by Bendell and Gebhart [1].

4. E. Grande. Analysis and Conceptual Design of Practical Ice Water Heat Sink, USACRREL Special Report 221, 1975.

5. H.J. Merk. Appl. Scient. Res., (1954), Section A, Vol. 4, pp. 435-452.

6. R.S. Schechter and H.S. Isbin. AIChE J., (1958), pp. 81-89.

7. H. Schlichting. Boundary Layer Theory, Pergamon Press, London, 1955.

8. J. Stubstad and W.F. Quinn. An Experimental Scaling Study of an Annular Flow Ice-Water Heat Sink, USACRREL Research Report 77-15, 1977.

9. A.G. Tkachev. Heat Exchange in Melting of Ice, Atomic Energy Commission Translation 3405, 1953.

10. C.R. Vanier and C. Tien. Effect of Maximum density and melting on Natural Convection Heat Transfer from a Vertical Plate, Ninth Natl. Heat Transfer Conference, AIChE preprint No. 3, 1967.

11. Y.C. Yen and C. Tien. J. Geophys. Research, (1962), pp. 3673-3678.

Yin-Chao Yen
Physical Sciences Branch
Research Division
U.S. Army Cold Regions Research and Engineering Laboratory
Hanover, NH 03755

Panel Discussion

exposed the differing realities of the mathematician and engineer. There was a communication problem, and most of the presentations showed little concern for the variety of backgrounds of the audience. The workshop aspect was not a success, and the modeling area was almost completely lacking. Thus, a number of speakers presented complex models with no explanation of their source. Technically the meeting was well organized, but the speakers should have been asked to clarify their modeling efforts. A bright spot was the paper of Fix, in which activities of modeling, analysis and computation all came together. As to the state of the art, most difficult problems remain unresolved. More experimental work is needed, and a greater appreciation of its importance is desirable. I believe that theoretical mathematical work will not have a significant impact on practical problems, and we (the engineers) will have to solve them with whatever tools are at our disposal. Because of its briefness, the meeting will have little impact on bridging gaps. What we really need is greater tolerance and respect between fields. As the editor of the *Journal of Heat Transfer*, I believe that publishing papers in each other's journals would be appropriate and useful.

SOLOMON: We are now open to questions and comments.

MEYER: I appreciate Professor Sparrow's view that the problems "will be solved." I would like to point out that numerical analysis is now being applied at the limits of knowledge, and such results as there are today in error analysis and correctness of algorithms cannot be easily extended. Certainly it is relevant to point at the example of finite element methods where effective application proceeded the recent work on error analysis and convergence by a number of years.

SPARROW: My point was not that mathematicians do not face a variety

PANEL DISCUSSION

The last event of the symposium was a panel discussion. The panel consisted of A. D. Solomon, Chairman, Professors Boley, Cannon, Meyer and Sparrow. The discussion began with each member addressing the following questions.

A) What was your impression of the symposium?

B) What is the current "state-of-the-art" on moving boundary problems?

C) What are your impressions of the problem of "bridging the gaps" between, e.g. the engineers, the numerical analysts, and the theoreticians working in this area?

The discussion was then opened to questions and comments from the floor.

BOLEY: Workers in this area may be characterized according to their interests in theory, methodology of solving problems, engineering, and modeling of physical problems. A significant result of the meeting was to establish a fragile bridge between the formerly disjoint groups of engineers and mathematical theoreticians; on the other hand, the modelers at present only interact with the engineers. A significant flaw of the meeting was the absence of modelers. At present the state of the art differs from area to area; for example, crystal growth is at the early stages of analysis, while casting practice is very advanced. Regarding "bridging gaps," the role of mathematical theory is restricted because it interacts principally with methodology, but only very weakly with engineering. A crude and obviously unfair caricature is that the mathematician does not care to

ISBN 0-12-757350-X

gain a good understanding of the underlying physical problem; rather, he wishes to generalize the problem so that it can be applied to many problems, of which his own understanding is poor. If successful this way is really impressive. However, the realities of the difficult problems facing us often render such general approaches of little value. The conference has succeeded in providing an opportunity for osculation between engineers and mathematicians and, based on a healthy attitude, a kiss is a good beginning.

CANNON: The symposium was well organized and served the important purpose of bringing together workers in the field and in promoting an exchange of information. The published proceedings will have but minor impact on research. The present state of the art is that for one-phase problems the theory is essentially complete while many aspects are also complete for the 2-phase case. A vast literature exists on Navier-Stokes equations and parabolic systems, while little work has been done on degenerate problems. The gap between groups is a natural result of the development of science and technology over the last century, with few if any people knowledgeable in broad areas of science. Thus the days of giants like Cauchy, Dirichlet, Gauss, Lamb and Gibbs ended in the late nineteenth century. The last 70 years have seen the rise of the large professional societies, the information explosion and computers. Our educational system is failing with a lack of successful teaching of basic information. We need a B.S. of 5-6 years with strong emphasis on rhetoric, composition, mathematics, physics, chemistry, biology, computing, logic and the humanities, and no specialization until a high level of achievement is reached in most of these things. As to working together, the process of joint work, like marriage, is

based on the need of help, effective communication and fruitfulness. Shotgun-like cooperation in which research teams are organized and strongly managed will generally work only when the team is relatively homogeneous and cooperative. Such efforts often fail as well. One hoped-for result of the meeting is a future scientific "marriage."

MEYER: The workshop aspect of the meeting was lost, and in fact, the meeting consisted of several symposia on different areas. The meeting should have had fewer papers and speakers and longer talks. The present state of the art is that many analytical aspects of moving boundary problems are well in hand; however, there has been little actual computation of situations involving complicated geometries or 3-dimensions. Similarly little work has been done on coupled problems involving combined heat and mass transfer. An example of fruitful work is that described in the talk of George Fix. Regarding the "bridging of gaps" I feel that the pure mathematician has been unfairly picked on in many circles at the meeting. I was struck by the lack of understanding of what a mathematician does, and by the feeling that his work must be "useful" or it is not good. I believe that the mathematician should be permitted to be considered as a philosopher and should not be so strongly required to "justify" his interests. The meeting will not create a "marriage" between areas. On the other hand, there can be an increase in the awareness, appreciation and contact that we maintain between areas. One possible way of encouraging this might be for individuals from the various groups to attend each other's professional meetings.

SPARROW: I consider the meeting to be somewhat of a dream sequence with many disconnected parts. I was surprised by many of the attitudes and orientations which expressed and

of difficult problems, but that many are not even interested in the relevance of their work.

MEYER: It is difficult to overcome attitudes.

BOLEY: I agree with Professor Sparrow that the engineer will "solve" his problems with or without external help. The theoretician must, however, pay attention to the existence of the engineering area; similarly, a genuine understanding is needed between the engineer and the mathematician. I would not criticize the remarks of Professor Meyer concerning the mathematician as a philosopher except for my experience that he is hungry for applications. I would say that he should stop fence straddling. I do agree with Professor Sparrow that more tailoring of the talks to the audience would have been desirable. Concerning the remarks of Professor Cannon, I would take exception to the claim that "well-roundedness" was confined to people of the last century. As an example, I might mention Theodore Von Karman. The criticism of the meeting was well taken, but at least we have had a kiss if not a marriage.

SPARROW: I might again stress the importance that I attach to the modeling process.

CHANDRA: From the beginning I was skeptical about the workshop aspect of the meeting. The workshop was not achieved. Similarly a meeting of this type should stress people who can give good presentations, rather than simply those in the forefront of research.

SPARROW: How many speakers here would be in the category?

CHANDRA: We had here a mixed bag of people with all the signs of an international conference. In this atmosphere a workshop was impossible. An additional problem was that speakers and topics changed too fast. Nevertheless some of the talks were good.

BOLEY: I would like to ask John Ockendon if his last Oxford meeting encountered such problems.

OCKENDON: Yes, although we did restrict ourselves to smaller groups of speakers. The meeting did not "bridge gaps."

BOGGS: I would like to ask Professor Sparrow how a meeting stressing modeling could be organized.

SPARROW: The complexity of an answer to your question is seen in the contributed papers sessions of this meeting. The modeling procedure, with its unrelenting attempts to produce a meaningful, yet tractible, mathematical formulation should be exhibited.

HILL: I would have greatly enjoyed a workshop atmosphere. I believe that the requirement that expenses would be supported only for speakers should not have been made. Generally speaking, the mathematician does desire to know what are the physical problems, and we would all have gained from a workshop atmosphere.

SPARROW: You mean that you would like to observe and be a part of the modeling process.

HILL: Yes--iteratively with the other disciplines.

ODEN: Based on my own experience, I would take some exception to the assertion that the modeling and mathematical analysis processes are disjoint. Certainly in areas of elasticity and continuum mechanics a belief in the validity of a model is greatly aided by an existence theory for the model. In particular an existence theory is the only tool besides physical arguments and experiment for testing the validity of a theory. Hence, a meeting devoted to modeling must in part be devoted to work on existence. In addition, existence results often point the way to computational approaches, an example being provided by recent work on variational inequalities.

FLEISHMAN: I enjoyed and benefited very much from the meeting. In addition to learning about new problems, I met people whom I did not know earlier. As in all things there is room for improvement. The most significant flaw of the meeting was that its goals were not defined and hence it is open to some of the criticisms we have heard. Perhaps a future meeting should have no requirement on its participants to present papers and should restrict itself to perhaps a half dozen problem areas. If its purpose were to have a workshop, then it would be successful. I might add my belief that the mathematicians do want to hear of problems.

PARKE: One approach to "bridging gaps" might be to have people take their sabbaticals in departments which are in fields that are not their own. Thus I, an engineer, benefited greatly by spending a year in a physics department.

FIX: I believe that the engineering community does make use of results of mathematical analysis, which do "filter down" to them.

ROSE: There should have been more free time for discussion at the meeting.

CRYER: The lecture periods were too long. I believe that machinery exists for broadening the areas of effective mathematical application. There is certainly an iterative process going on, with the choice of a model often being determined by the ease and possibility of solving it.

MEYER: The modeler is often moved by his own background.

BOLEY: You mean, that determining what to ignore and what to consider in the process of modeling is a joint effort.

The discussion was then adjourned.

Author Index

Subject Index

G

H

I

T

Y

Z

A
B
C 8
D 9
E 0
F 1
G 2
H 3
I 4
J 5

MOVING BOUNDARY PROBLEMS

PANEL DISCUSSION

The last event of the symposium was a panel discussion. The panel consisted of A. D. Solomon, Chairman, Professors Boley, Cannon, Meyer and Sparrow. The discussion began with each member addressing the following questions.

A) What was your impression of the symposium?

B) What is the current "state-of-the-art" on moving boundary problems?

C) What are your impressions of the problem of "bridging the gaps" between, e.g. the engineers, the numerical analysts, and the theoreticians working in this area?

The discussion was then opened to questions and comments from the floor.

BOLEY: Workers in this area may be characterized according to their interests in theory, methodology of solving problems, engineering, and modeling of physical problems. A significant result of the meeting was to establish a fragile bridge between the formerly disjoint groups of engineers and mathematical theoreticians; on the other hand, the modelers at present only interact with the engineers. A significant flaw of the meeting was the absence of modelers. At present the state of the art differs from area to area; for example, crystal growth is at the early stages of analysis, while casting practice is very advanced. Regarding "bridging gaps," the role of mathematical theory is restricted because it interacts principally with methodology, but only very weakly with engineering. A crude and obviously unfair caricature is that the mathematician does not care to

ISBN 0-12-757350-X

gain a good understanding of the underlying physical problem; rather, he wishes to generalize the problem so that it can be applied to many problems, of which his own understanding is poor. If successful this way is really impressive. However, the realities of the difficult problems facing us often render such general approaches of little value. The conference has succeeded in providing an opportunity for osculation between engineers and mathematicians and, based on a healthy attitude, a kiss is a good beginning.

CANNON: The symposium was well organized and served the important purpose of bringing together workers in the field and in promoting an exchange of information. The published proceedings will have but minor impact on research. The present state of the art is that for one-phase problems the theory is essentially complete while many aspects are also complete for the 2-phase case. A vast literature exists on Navier-Stokes equations and parabolic systems, while little work has been done on degenerate problems. The gap between groups is a natural result of the development of science and technology over the last century, with few if any people knowledgeable in broad areas of science. Thus the days of giants like Cauchy, Dirichlet, Gauss, Lamb and Gibbs ended in the late nineteenth century. The last 70 years have seen the rise of the large professional societies, the information explosion and computers. Our educational system is failing with a lack of successful teaching of basic information. We need a B.S. of 5-6 years with strong emphasis on rhetoric, composition, mathematics, physics, chemistry, biology, computing, logic and the humanities, and no specialization until a high level of achievement is reached in most of these things. As to working together, the process of joint work, like marriage, is based on the need of help, effective communication and fruitfulness. Shotgun-like cooperation in which research teams are organized and strongly managed will generally work only when the team is relatively homogeneous and cooperative. Such efforts often fail as well. One hoped-for result of the meeting is a future scientific "marriage."

MEYER: The workshop aspect of the meeting was lost, and in fact, the meeting consisted of several symposia on different areas. The meeting should have had fewer papers and speakers and longer talks. The present state of the art is that many analytical aspects of moving boundary problems are well in hand; however, there has been little actual computation of situations involving complicated geometries or 3-dimensions. Similarly little work has been done on coupled problems involving combined heat and mass transfer. An example of fruitful work is that described in the talk of George Fix. Regarding the "bridging of gaps" I feel that the pure mathematician has been unfairly picked on in many circles at the meeting. I was struck by the lack of understanding of what a mathematician does, and by the feeling that his work must be "useful" or it is not good. I believe that the mathematician should be permitted to be considered as a philosopher and should not be so strongly required to "justify" his interests. The meeting will not create a "marriage" between areas. On the other hand, there can be an increase in the awareness, appreciation and contact that we maintain between areas. One possible way of encouraging this might be for individuals from the various groups to attend each other's professional meetings.

SPARROW: I consider the meeting to be somewhat of a dream sequence with many disconnected parts. I was surprised by many of the attitudes and orientations which expressed and exposed the differing realities of the mathematician and engineer. There was a communication problem, and most of the presentations showed little concern for the variety of backgrounds of the audience. The workshop aspect was not a success, and the modeling area was almost completely lacking. Thus, a number of speakers presented complex models with no explanation of their source. Technically the meeting was well organized, but the speakers should have been asked to clarify their modeling efforts. A bright spot was the paper of Fix, in which activities of modeling, analysis and computation all came together. As to the state of the art, most difficult problems remain unresolved. More experimental work is needed, and a greater appreciation of its importance is desirable. I believe that theoretical mathematical work will not have a significant impact on practical problems, and we (the engineers) will have to solve them with whatever tools are at our disposal. Because of its briefness, the meeting will have little impact on bridging gaps. What we really need is greater tolerance and respect between fields. As the editor of the *Journal of Heat Transfer*, I believe that publishing papers in each other's journals would be appropriate and useful.

SOLOMON: We are now open to questions and comments.

MEYER: I appreciate Professor Sparrow's view that the problems "will be solved." I would like to point out that numerical analysis is now being applied at the limits of knowledge, and such results as there are today in error analysis and correctness of algorithms cannot be easily extended. Certainly it is relevant to point at the example of finite element methods where effective application proceeded the recent work on error analysis and convergence by a number of years.

SPARROW: My point was not that mathematicians do not face a variety